The Archaeology of Iron Production on Ethiopia

Temesgen Burka

The Archaeology of Iron Production on Ethiopia

Ethnoarchaeometallurgical Case Studies from Wollega

Galda Verlag

Die Deutsche Bibliothek - CIP-Einheitsaufnahme
Die Deutsche Bibliothek verzeichnet diese Publikation in der Deutschen Nationalbibliografie; detaillierte bibliografische Daten sind im Internet über http://dnb.ddb.de abrufbar.
ISBN 978-3-941267-00-8

ISBN 978-3-941267-00-8

ACKNOWLEDGEMENTS

I thank Dr. Kassaye Begashaw for giving me the opportunity to share experience with Professor Randi Haaland who brought with her lively thought in the study of African ethnoarchaeometallurgy. Dr. Kassaye should be credited for taking the burdensome responsibility for leading the Archaeology Unit from its sole MA program into the full-fledged Department of Archaeology and Heritage Management.

It has been a great honor for me to have the supervision and advice of Professor Rodolfo Fattovich, who was also an asset for the actualization of our Department at AAU. He has been a great inspiration for developing the Third Model in discussing the archaeology of iron production in Ethiopia by situating the Wollega ethnographic record in archaeological context. In addition, similar gratitude should be owed to Professor Claudio Giardino, the co-supervisor, whose advice went beyond reading the dissertation and commenting to one of opening up opportunities for presenting my findings in international conferences, and co-publishing. His connections have enabled me building an experimental smelting setting at Antiquitates (March-April, 2011) where I did a close examination of variables in traditional iron smelting.

Many have contributed their respective resources for the success of this research. There is however nothing that equals the tumtuu and buuftuu of Abee-Dongoro and Aira-Gulliso. I have to thank colleagues at AAU who shared their precious time for reading the manuscript and suggesting their comments. My thanks in this regard go to Fedhaa, Belete Bizuneh, and David Chapple. I also thank the AAU management particularly Professor Andreas Eshete for top-upping my study at "L'Orientale," which otherwise could not have been easy with meager scholarship from Italian Ministry of Foreign Affairs. In connection, the role of Professor Alessandro Triulzi, PhD coordinator is an imaginable, which was able to fill-in the financial constraints created due to irregularity in scholarship payment. In this regard, the administrative issues in my scholarship could not have been simple without close attendance of Dr. Andrea Manzo and fatherly coach by Professor Yacob Beyene. Other colleagues at "L'Orientale" including Drs. Luisa Sernicola, Antonella Brita, as well as those in the administration and library of Department of African and Arabian Studies require deep appreciation. I should say many thanks to others including Chernet Waqweyya, Tekalign W/mariam, and master smelters/smiths from Malkee, Yamalagii, Abayyanii, Laaloo, Innagamdii, Innawangii, Innawarii, Innabalxii, and the like. Obbo Chernet Waqweyya is particularly important in helping me access to the source of the tradition in northeast Wollega. He did not only accompany me as field assistant but he was also instrumental in deciphering the messages kept in various traditions regarding the origin of tumtuu artisans, their essential roles in the society as well as circumstances leading to their current position.

Finally, the biggest debt of gratitude is owed to my wife (Banchamlak Mehari) and my children (Bilisuma, Gada, and Sifan) for putting up with me. I am indebted for you all doing without daddy and husband for many years and without complaining much about it. Your love, encouragement, advice, and confidence in me have sustained me throughout this experience and are a means of great success. Thanks and I love you very much! Chumba, I owe you!

TABLE OF CONTENTS

LIST OF CAPTIONS

PREFACE

One of the most spectacular features of Ethiopia and the Horn is its strategic geographical location vis-à-vis its close (the Horn of African nations) and distant neighbors northwards to Egypt and eastwards to the Arabian Peninsula. Ethiopia's geography, like its people, has a long and dynamic history, resulting in a scenic landscape of variety and splendor (Levine 1974). Topographically, Ethiopia can be divided into three major zones: the cool highlands (baddaa) where the annual average temperature is about 16 degrees centigrade, the intermediate zone (badda-daree) where most of the settled population lives, and the hot valleys and plains (gammojjii), which are dependent on the desert conditions (Megerssa 1994; Wolde-Mariam 1972).

Ethiopia contains four major rivers systems (Wolde-Mariam 1972) namely the Takkaze, the Abbay/Mormor (Blue Nile), and the Baro, originating in parts of Wollega, and flowing westward into the Nile. The Abbay/Mormor is the most famous, deriving its source from Gishe Abbay in the highlands of Gojjam and further augmented with multitudes of tributaries such as Jamaa, Gudar,

Fincaa'aa, Amartii, Nashe, Hangar-Dhidheessa, and Dabus from Shoa and Wollega. The second group consists of the Ganale and the Wabe Shabale river system. These two rivers flow towards the Indian Ocean after providing fresh water and alluvium to the hot and dry lands of the Somali inhabiting of Ethiopia and Somalia. The Gibe, (also known as Omo in its lower course) originating from the cool highland part of Wollega, flows through the southwestern highlands, with the Turkana, also known as Lake Rudolph, on the Ethio-Kenya border as its destination. The Awash sets off from the highlands of Cillimoo forest in west Shoa zone, close to the town of Gincii, streams across the Rift Valley, and vanishes in its northeastern sands at Lake Abbe (see Wolde-Mariam 1972; also Hassen 1990; Zewde 1991 and Levine 1974; Waqweyya 1988).

This stunning physical setting of the Ethiopian region and the Horn has offered both advantages and disadvantages (Burka 2006). It has particularly affected the attempt to reconstruct its culture-historic past. On the one hand, due to its strategic position connecting continents, it can be assumed that the region has benefited from millennia long cultural contact with its neighbors (Fattovich 1996), the result of which is rich and diversified cultural developments. On the other hand, however, the exposure of the region to the outside has led to assertions, by diffusionists, that various cultural developments in the region ultimately owe their origin to external sources (e.g. Van Beek 1967).

In this dissertation, I argue that the temptation to ascribe ancient Ethiopian cultural developments to outside influences resulted from the focus of archaeological research in Ethiopia (e.g. Aksum) on elite monumental and architectural objects and features (see Simoons 1965:9). The contents of royal cemeteries, for instance, which provide exotic goods, have led researchers to speculate that ancient developments in Ethiopian and the Horn were replicas of processes or developments from regions across the Red Sea in South Arabia or even the Nile valley. It suffered from structural problem.

Moreover, the focus of Ethiopian archaeological research on elite residential, political, as well as symbolic sites has prevented the systematic investigation of the settlement areas of the common people. Consequently, the study of cultural traits such as the archaeology of crafts namely metallurgy has remained in the backwaters of Ethiopian archaeological enquiry. In this work, I argue that the dearth of archaeological record for Ethiopia's ancient metallurgical history (Haaland 1985; Phillipson 1993; Todd 1985) is not so much because of the lack of the cultural trait but because of the absence of research focus on this particular theme (Burka 2006). In this regard, recent ethnographic and ethno-historical studies on traditional metallurgy in some parts of Ethiopia are evidence of the possible breadth and depth of this cultural trait in various parts of the country. The Wollega ethnoarchaeometallurgical material, with which this dissertation deals, further strengthens this argument.

My interest in an ethnoarchaeometallurgical[1] study of metallurgy in Wollega goes back to late 2004 when I took up the subject as a topic for my M. Phil thesis at the University of Bergen (Burka 2006). Accordingly, various factors led me to focus on Wollega in this dissertation. Firstly, the region is one of the least researched areas in Ethiopia's historical, anthropological, linguistics, and especially archaeological studies. In addition, I have personal reasons for picking Wollega as my geographical focus of research. I was born and grown up in a district close to the Blue Nile in northeast Wollega. I grew up listening to local stories, myths, and traditions including that about smiths. As a result, there is always a personal move. Therefore, I realized that if I chose Wollega as a study area, I would not face a number of constraints such as the 'other' would in an attempt to access to and obtain data from the region. In other words, I realized the advantages I would get as an insider in dealing with the intricacies in the ethnoarchaeometallurgical study. This is the first work by an insider on ethnoarchaeometallurgy in Ethiopia.

In addition, during my MA research at Addis Ababa University, on Ethiopian Museums in 2003, one of the sites I analyzed was the Wollega museum. This museum juxtaposes collections ranging from natural history collections including various tree species, animal products to handicraft technology that promote local knowledge. This information fascinated me. I was especially intrigued by displays particularly those on metallurgy with ores, the smith's working tools and the various metal objects such as spear points and axes claiming to have local origin. In addition to information from natives to southwest and northeast Wollega, (e.g. Chernet Waqweyyaa-historian and Bulii Ejjetaa- anthropologist), the literature review I made in the autumn of 2004 on the region boosted my confidence to start from scratch, this regional study. The archaeology of iron production in Wollega is constrained largely by lack of previous works on this or other archaeological topics in the region. As a result, no temporal sequence is available for prehistory of Wollega in particular.

---

1 My use of the term ethnoarchaeometallurgy instead of ethnoarchaeology is deliberate in that the purpose is to indicate my emphasis on the archaeology of iron production. On the other hand, however, the method I employed is an ethnoarchaeological one and I also used both terms in this text.

I carried out field trips in different parts of Wollega. Finally, I conducted, a primary ethnoarchaeometallurgical re-enactment on iron smelting in Abee Dongoroo in 2005. Then, further enquiry into regional history resulted in another fieldwork in Aira-Gulliso region in the year 2007 that helped to reconstruct local iron smelting practices. The Aira-Gulliso project has been an additional impetus for developing the PhD project proposal that led finally to the present dissertation. Although, the major focus of the present research has been on Aira-Gulliso, southwest Wollega, iron-smelting practice in Abee Dongoroo has also been integrated, as it was a unique opportunity to document and analyze technological variability in metallurgy even among groups who claim to have common lineage origin. Abee-Dongoroo is particularly informative on the nature of social stratification in the Oromo society of Wollega.

The dissertation is organized in seven chapters. The first chapter offers a theoretical and methodological framework for the study of the archaeology of African metallurgy. Two theoretical issues have framed the study of African culture-history particularly that deals with the introduction of iron technology. These are single origin and multiple origins debates. In addition, the chapter presents the various models accounted for the introduction of iron metallurgy in Ethiopia and the Horn vis-à-vis the surrounding regional setting, in spite of lack of research focus. Nevertheless, this chapter while presenting both antagonistic positions (at continental and sub-continental level) argues in favor of inclusive standpoints, which I labeled as the Third Model. Ethiopian (as well as Wollega) geographic position and culture-historic past favors the search for middle grounds. Methodologically, ethnoarchaeology and the chaîne opératoire frameworks were regarded as rails upon which the reconstruction of the operational sequences that resulted in the production of certain metal tools before they enter into the archaeometallurgical record were founded. Chapter I also discusses the gap of research in Ethiopian archaeometallurgy in general and Wollega in particular. In addition, it outlines the research with objectives, questions, and data acquisition techniques.

Chapter Two provides a description of the selected study areas-the peopling and the environment. The chapter deals with two geographically separated districts located about 200 kilometers from each other in different water shades in Wollega. One of the study areas is located in Abee-Dongoro in northeast Wollega while the second site is found in Aira Gulliso, southwest Wollega. The chapter outlines the nature of social stratification among the Oromo of Wollega enabling us locate the ranks given to different occupational identities. By presenting traditions and the nature of commensality in the society, the chapter attempts to distinguish between the class of smelters and the smiths. One of the major themes the chapter addresses is the fact that the artisans in both sites claim a common lineage origin. Yet, despite the claim of common ancestral or lineage origin, they have employed different technologies of iron smelting. In addition, the chapter discusses the natural or environmental setting of both study areas. It outlines that the geographical position of Wollega connecting various regions and cultures would encourage endorsing the Third Model. By concentrating on the local geology and geography, it attempts to situate the culture of iron metallurgy in the kind of natural resources that might have enabled it.

Chapter Three is a follow-up of chapter one. In this chapter, the chaîne opératoire, both as theoretical framework and methodological tool is discussed to test its applicability in the study of the archaeology of iron metallurgy particularly with ethnoarchaeological approaches. The chapter discusses the various stages through which the objects pass all the way from nature to culture encompassing various levels of relationships. In this regard, it thoroughly examines the research model for conducting ethnoarchaeological studies of iron metallurgy in Wollega. It is where I attempted to show that ethnographic study informed by the chaîne opératoire model is instrumental for understanding how humans in the past interacted among themselves and with their environment in the medium of things. In whole, the chapter forms the foundation upon which Chapter five that deals with the case study is displayed.

Chapter Four, deals with the survey and documentation of archaeometallurgical sites in northeast and southwest Wollega. As with the ethnoarchaeometallurgical/ethnoarchaeological portion of the dissertation (Chapter Five), this chapter also focuses mainly on metallurgical sites in Aira-Gulliso district of southwest Wollega. I focus on Aira area in particular not only due to the high concentration of iron production sites and widespread exploitation of the environment in the district, but also because the most ubiquitous metallurgy sites are located inside the town. This calls for special attention from heritage management point of view. The fact that the town is under plan revision with road and house construction activities on hitherto fringe areas strengthens my decision to give much space on metallurgy sites in the environs of the town. In this case, various vestiges of archaeometallurgy sites were systematically surveyed and recorded. Future comprehensive work would enable us enlighten with the depth of regional culture-history through archaeometallurgy.

Chapter Five forms the core of the Dissertation. Two case studies of ethnoarchaeometallurgical data, one from northeast Wollega, and the other from southwest region of the same region, were taken for specific study and modeling interpretation of archaeometallurgical record. I found out that through reconstruction of formerly practiced traditional metallurgy, it is possible to revitalize indigenous knowledge as well as create a suitable ground for interpretation of archaeometallurgical vestiges. The social and environmental challenges and opportunities offered to artisans could be able to distinguish under this chapter. The chapter enables to theorize how humans respond to the various constraints (cultural and environmental) during transformation of things.

The sociality of iron is dealt with under Chapter Six. Under this chapter, I sought to redress many of the limitations in the study of iron metallurgy, which are caused by focusing on the physical aspects of the knowledge. It is argued here that the study of the origin and status of artisans is a significant contribution to the reconstruction of a history of metallurgy. The chapter deals with various models for interpretation of the origin of the so-called marginalized artisans in the Ethiopian region. In addition, it presents the traditions of the mainstream society and the smiths on the myth of their origin and historical circumstances attributed to the origin of 'marginalization' of the tumtuu of Wollega (Burka 2009). Finally, the ritual role of iron objects among the Oromo of Wollega, particularly the symbolic role of metal tools, is presented in connection with traditional ritual sites, the Qaalluu.

Chapter Seven is the synthesis of the whole work of the dissertation. It puts together the state of the art in the discourse of the origin and distribution of the knowledge. The chapter puts the technological and socio-cultural aspects of iron metallurgy in Wollega in its wider setting. It is under this chapter that the applicability of ethnographic data for archaeological interpretation of past ways of life is scrutinized. The various aspects of ethnoarchaeometallurgical evidence in Wollega (in both northeast and southwest) offered unrivaled opportunity for archaeometallurgists far detached from the reality of the present. Furthermore, I have also offered, under this chapter, scenarios for the introduction of the knowledge of metallurgy in Wollega. As part of the scenarios for the origin (based on the Third Model), it raises a challenging question of whether migrationist approach (a history of Oromo population movement in 16th century) is explaining the means for the origin of metallurgy in Wollega. Moreover, it presents the challenge on the single origin hypothesis of the knowledge, as people of close lineages possessed radically different technologies of iron metallurgy. It has, as a result, shown that models such as population migrations are not sufficient to answer questions such as the spread of certain cultural trait exemplified here by iron metallurgy. In other words, the Wollega material, as synthesized under this chapter has strengthened the assumption that the region might have developed technologies in order to meet local specific requirements.

CHAPTER ONE

1.0. GENERAL PRESENTATION

1.1. ETHIOPIA AND ITS NEIGHBORS: PERSPECTIVES ON THE ORIGIN OF METALLURGY IN THE REGION

...Each race is endowed with particular talents..., the freedom, and independence of each. In the music of the universe, each gives a different sound, but all were necessary to the grand symphony. There are several sounds not yet brought on and the feeblest of all is that hither to be produced by the Negro [sic], but only he can furnish it...

Adapted from A. Holl (1997's: 59 quotation of E.W. Blyden, a pan-Africanist)

1.1.1. THEORETICAL FRAMEWORK

Eurocentric and Afrocentric perspectives characterize the archaeological study of iron production in Africa. Central to these approaches is the question of the origin of the knowledge. Their question is whether iron metallurgy originated in the continent or introduced from outside. The discussion sets apart the continent into sub-Saharan Africa on the one hand and northern African and the Nile valley on the other. Specifically, the sub-Saharan Africa has served as the battleground between the two antagonistic sides (see Childs and Herbert 2005; Childs and Killick 1993; Herbert 2001; Killick 1999, 2009; Mapunda 2000; Mitchell 2005; Schmidt and Childs 1985; Schmidt 1997, 2001; also Burka 2006). Some of the specific questions this literature addresses include; - Where did Africa (sub-Saharan Africa) obtain the knowledge of iron metallurgy? Which routes were the likely alternative conduits for funneling the knowledge into the sub-continent? The other intriguing issue was whether Africans were able to pass directly from Stone Age cultures to Iron Age on their own.

These questions are not, however, without merit. The challenge must be measured in the degree of effort and time allotted to the discussion of its origin/source while disregarding many other things accompanying the introduction of the knowledge of iron metallurgy. On the other hand, the question of origin is common in various aspects of archaeological enquiry. Such questions including the origin of certain cultural trait, whether the trait belongs to closer or distant societies as well as the question of the identity of the owners or the carriers of the knowledge etc have always been integral parts in the study of any prehistoric or historic past.

Nevertheless, a great deal of focus on the question of origin, leaving many things aside, in the study of African metallurgy has resulted in adverse consequences. Part of the consequences of such allure with the question of origin, according to its critics, is that it has neglected the contributions of the technology once it appeared among the different peoples in the continent (see Mapunda 2000; Schmidt 1997, 2001). In spite of the origin of the knowledge, the themes, which could have shed more light on the study and provided a wealth of information, are many. An important focus could have been exploring the constructive or destructive roles iron metallurgy played once it was integrated into the life of the people. In other words, how Africans responded to its introduction in their respective ecological niche could have provided insight into diversity in the continent. In addition, did, for instance, the introduction of the knowledge accelerate development by increasing efficiency in bush clearing, food production, increased population size and resulted in the emergence of more complex systems (see De Barros 2000)? These and other similar aspects of the study have been left in the backwaters of the discourse in sub-Saharan Africa due to obsession with its origins.

Furthermore, regardless of the scientific basis of the question of origin, it has led, when it comes to metallurgy, largely to a conclusion that supports an outside origin-**a Cairo to Cape schema**- of the knowledge (see Arkell 1966, 1968; Killick 2009) as opposed to allowing multivocality (see Blyden cited in Holl 1997:59). According to many Africanist scholars, the outside origin perspective is rooted in the colonial baggage aimed at denying Africans' innovative past (see Chirikure 2005; Diop 1974, 1981; Mapunda 1997; Schmidt 1997). The attempt, by diffusionists, to disengage the source of the knowledge from culture-history of the continent has further complicated the discourse on African archaeometallurgy by breeding antagonistic perspectives especially from African nationalists.

The protagonists of outside origin envisage four routes as conduits for the introduction of iron technology in sub-Saharan Africa (see Miller and Van der Merwe 1994; Van der Merwe 1980). These are the Maghreb in Morocco, the Gulf of Sirte in Libya, Egypt, and the East African maritime corridors (see Arkell 1966, 1968; Killick 2009; Miller and Van Der Merwe 1994; Tylecote 1975; Van Der Merwe 1980; Wainwright 1942a, 1942b, and for comments Mapunda 1997). For the purpose of clarity and the regional scope of the study, the following discussion is delimited to northeast Africa-Egypt and its immediate southeastern neighbors i.e. Ethiopia and the Horn.

Partly due to its geographic proximity, Egypt has been regarded as one of the major conduits through which Near Eastern civilizations (e.g. from Levant) have trickled to the rest of Africa (see Arkell 1968; Killick 2009:411; for further reading Mapunda 1997; Shinnie 1967, 1985; Trigger 1969; Wainwright 1942a, 1942b). However, lack of archaeological evidence (probably due to lack of focus on this particular topic) for earlier dates in Egypt than many sites in other parts of African regions to its south has led to question the reliability of this assumption (Mapunda 1997; Trigger 1969). Mapunda (1997), who patched up evidence from the Horn in particular, posits that the evidence is against Egypt than in favor of it. In contrast, it is however, argued that that the dearth of evidence on Egyptian ancient archaeometallurgy could be attributed to lack of focus of study on this topic (see Killick 2009) than lack of the cultural trait.

Thus, the outside origin hypothesis for African civilization has complicated the study of the archaeology of African metallurgy not only by marginalizing its impacts on the society but also by fueling antagonism from nationalist archaeologists (see Holl 1997; Killick 1999; Mapunda 2000; Okafor 1993; Shepherd 2002; Trigger 1969, 1984, 1989). As a result, the archaeology of iron metallurgy in Africa (not so really in Ethiopia and the Horn) has become a battlefield between diffusionists and nationalists (see Childs and Herbert 2005; Shepherd 2002; Killick 2009). The difference between the two lies in the interpretation of the same archaeological record. In addition, the discourse regarding African metallurgy is a reflection of an ongoing politics in African archaeology in general. Nick Shepherd (2002:189) outlines the antiquity and diversity of politics in African archaeological research in the following manner:

> ...The politics of archaeology in Africa has been no less marked by variety. From the savants who accompanied Napoleon's army of conquest in Egypt to the fiercely nationalist archaeologists of Africa's independence, from the overtly colonialist agendas of early researchers to contemporary debates over human origins and racial diversification, and from projects of excavation aimed at uncovering the glories of indigenous states to those aimed at recalling the horrors of slavery, archaeologists have been implicated directly in political struggles and debates...

The outside origin hypothesis, which critics attributed it to be fused with colonialist perspective towards the African past, has led to the rise of antagonistic attitude primarily from African nationalists working in other fields than archaeology. The opposing standpoint (e.g. Cheikh Anta Diop 1974, 1981) is rooted in Afrocentric outlook, which designates the previous discourse as colonialist, Eurocentric and diffusionist (for detail see Chirikure 2005; Holl 1997; Trigger 1984).

Afrocentrists claim that the knowledge of iron metallurgy originated within Africa. Specifically, Egypt and ancient Egyptian civilization are described as black African (e.g. Diop 1968, 1974) from whence the knowledge spread not only to the continent but also (Diop 1981; Holl 1997) to Europe. Cheikh Anta Diop has championed the African origin standpoint. Augustine Holl (1997:59-60) holds that Cheikh Anta Diop '...saw the whole body of scientific interpretation of Africa's past as a tool used to enforce colonial dominance, since it deprived Africans of an aptitude for cultural achievement.'

Consequently, in the 1950s, Diop envisioned an independent and federally united Africa and a black African origin of humanity. In order to accomplish his worldview of Africa, Diop saw the need to revitalize African rich past in three steps. First, he decided to establish ancient Egypt as cradle of the African people. Then, he wanted to demonstrate the presence of links between Pharaonic Egypt and other African peoples and linguistic groups. Finally, he envisaged to single out Pharaonic Egypt as the cradle of civilization and center of dispersal not only within the continent but also to the western world particularly influencing ancient Greece (Holl 1997:60) from which other Europeans inherited various traits of African civilization. In other words, Cheikh Anta Diop saw Egypt as an epicenter of ancient civilization for the rest of African regions and the rest of the World.

Ironically, despite its strong nationalist root, the Afrocentirc view has suffered from the same handicaps of 'diffusionism' and 'migrationism' as had the advocates of outside origin (*cf. chapter VII*; see also Sinclair *et al* 1993:2). The only difference lies in locating the center of dispersal- inside *vis-à-vis* outside the continent. Thus, proponents of the

African origin for ancient civilizations have failed to propose an alternative model for the purpose of scientific discourse. Their contribution was to accentuate the rebuttal of the colonialist outlook of African prehistoric past, which labels it as 'backward' and 'uncivilized' (see Trigger1994). Theirs was part of African nationalist struggle and a call for revival of 'African prestige' that is claimed to have existed before the continent was subjugated to the servitude of European colonial domination (see Chirikure 2005; Holl 1997; Shepherd 2002; Trigger 1994, 2003). The reaction was against the colonial view that represented Africa before colonial era as 'dark' and 'backward'. It was an attempt to show Africa's 'bright past' before its growth was deflated by the conquest and occupation of white colonialists (see Chirikure 2005; Shepherd 2002). In other words, both protagonists were not arguing in favor of various African societies' capabilities to respond to their respective environments according to local requirements (see Holl 2008; Mapunda 2000; Schmidt 1997, 2001; Schmidt and Mapunda 1997; Shepherd 2002). As for the 'nationalists', they rather went a long way looking for another center of origin in Africa for the origin of iron metallurgy (see Burka 2006; Holl 1997, 2008; Sinclair *et al* 1993; cf. chapter VII) i.e. Egypt.

Thus, the Afrocentrists as well as their opponents have not considered Africa's enormous diversity (e.g. historical, cultural, socio-economic, political, and environmental). Such misrepresentation of African past is seen from attempting to homogenize their cultures as in the discourse the so-called African 'bloomery' iron smelting (see Schmidt 2001). Consequently, it can be challenged that as far as sub-Saharan Africa is concerned, singling out Egypt as the source or conduit of the knowledge of metallurgy makes no serious difference if the sub-continent finds itself on the receiving side. For the 'receiver' sub-Sahara African society, an object or idea from Egypt is as foreign as the one from across the Red and Mediterranean seas. According to critics, both antagonistic viewpoints have resulted from the sides' irrational racist color attitude and personal interests than any justice to scientific inquiry (see Trigger 1994 for paradigms in Sudan's archaeology).

Thus, nationalists have primarily come far way to tackle the colonialist negative outlook towards Africa's prehistoric and historic past (see Chirikure 2005; Holl 1997; Mapunda 1997, 2000; Schmidt 1997) which depicts it as 'uncivilized and backward'. Conversely, they have just been bogged down by the debates about origin. Accordingly, they failed to contribute much to the investigation of other aspects of African achievements because of and after the introduction of iron metallurgy.

The other aspect of the argument surrounding the history of African iron production is concerned with the technological aspect of the knowledge. This has also stemmed from the argument over the origin of the technology and the search for types. In order to substantiate the outside origin of the knowledge, proponents of the single origin thesis list the physical features and the observable and tangible aspects of iron metallurgy. In other words, the technological apparatuses of metallurgy (e.g. technical ceramics) were used as elements for locating sites and tracing routes of spread (e.g. Cline 1937; Shinnie 1985; Sutton 1985; Todd and Charles 1978b; Trigger 1969; Wainwright 1942a, 1942b). Scholars have criticized this approach for separating and/or treating the technology of African iron production out of the cultural milieu in which it is operative (Burka 2006; Mapunda 2000; Schmidt 1997, 2001; Schmidt and Childs 1985; Schmidt and Mapunda 1997).

The other shortcoming of the technological/the physical attributes/ approach in the study of African metallurgy is its departure from the nature of the study in the West (e.g. Europe) where methodologies from physical sciences were used. Whatever the justification for its application, the study of African metallurgy stands distinct from approaches developed in the West. Its focus is on ethno-historical and ethnoarchaeological means of addressing the problem. In other words, it lacks tools for mineralogical and chemical analysis of finds (e.g. Burka 2006, 2008; Haaland 2004a, b; Haaland *et al* 2002; Haaland *et al* 2004a, b; Haaland and Haaland 2007, 2008; Herbert 1993; Schmidt 1997; 2001; Schmidt and Childs 1985). There are, however, some exceptions (e.g. see Abdu and Gordon 2004; Chirikure 2005; Killick 1999; Miller 2002; Todd and Charles 1978b).

Nevertheless, studies that made the question of origin central to their discourse have apparently drawn the mental map of the spread of the cultural trait in the continent of Africa from north to south-from Cairo to Cape (see Killick 2009). Even within the 'diffusionist' camp, there is no south-north direction. This implies that the earliest dates should be sought in the north while the southern parts of the continent are on the receiving end. Such expansion was put into a 'cultural package' carried out by the Bantu speaking agriculturalists for the southern African regions (see Chirikure 2005; Killick 2009; Phillipson 2005; Schmidt 1997; Sinclair *et al* 1993; cf. chapter VII). In contrast to the outside/single origin assumption, an independent/multiple origin of the knowledge of metallurgy in sub-Saharan Africa has now become an alternative means to the discourse.

According to nationalist archaeologists, one of the challenges for the outside/single origin standpoint is the presence of various metallurgical technologies in different parts of the continent. Even worse, settlements close to one another or communities with close affinities possessed separate technologies (see Burka 2006, 2008, 2009). Close examinations of the technology reveal the complexity of the debates on the nature of African metallurgy. For instance, there must be an explanation for the presence of chimney/natural draught furnaces in one region and forced draught ones in others (e.g. Barndon 2004; Childs and Killick 1993; Cline 1937; Schmidt and Childs 1985). It is clear also that furnaces with superstructures are dominant in one region and underground structures are used in many other localities (e.g. Burka 2006, 2008; Sutton 1985; Tylecote 1975, 1980). Variability in bellows is another challenging problem. In one region, it is common to find bellows made of bags whereas pot/drum bellows are the technological preferences in neighboring regions. There is also variability in terms of tapping mechanism or separating the impurities from metallic iron (e.g. Schmidt 1997; 2001; Schmidt and Avery 1978). Whereas in some regions slag-tapping furnaces are used, in others slag-pit furnaces or furnaces without slag-tapping mechanism are employed. In addition, whereas many different types and sizes of tuyères are used in some regions, only few can be observed in the neighboring ones (see Burka 2006, 2008; Haberland 1959; Haaland 2004a, 2004b; Haaland et al 2000, Huntingford 1955). Moreover, the depth of and direction of penetration by the tuyères varies from place to place. In some (e.g. Ethiopia), the penetration is at the base of the furnace while it is from the top (vertical) in other regions (e.g. Burka 2006, 2008; Todd 1985; Todd and Charles 1978b). To put it simply, how is such a huge diversity explained by a single/northern origin of iron metallurgy in sub-Saharan Africa?

However, proponents of the outside origin/diffusionists still forward a number of explanations against challenges posed by advocates of multiple/independent origin perspective. One explanation is attributed to the introduction of the knowledge of metallurgy into Africa in different stages of its evolution (see Hall 1987). Thus, it implies that Africans might have received the knowledge of metallurgy at different levels of technological advancement from outside the continent. This explanation, however, might have originated from a colonialist standpoint that denigrates innovative capabilities of Africans (see Alpern 2005; Burka 2006; Chirikure 2005; Holl 2008).

On the other hand, the issue of African indigenous development of metallurgy or the challenge to pass to Iron Age directly from Stone Age is more complicated by the concept known as ***pyrotechnology*** (see Muhly 1980; Pliener 1980, 2000; Tylecote 1980; Van der Merwe 1980). The concept is connected to long period of experimentation with fire including previous experiences with smelting of soft metals such as copper, a prerequisite for iron metallurgy (see Killick 1999, 2004; Miller and Van der Merewe 1994; Phillipson 1993, 2005) when iron ore is used as a flux.

Thus, the knowledge of iron metallurgy could only be developed in stages namely through previous work experiences on softer metals such as copper and bronze, which Africans seem to have skipped (see Phillipson 2005). During such experiences where iron oxide is used as a flux, for instance, for copper smelting, an experienced smelter is said to have identified the iron fragments, made some objects from them, and gradually developed the technology of iron metallurgy (Pliener 1980, 2000; Van der Merwe 1980). On the other hand, sub-Saharan Africans are said to have passed from Stone Age directly to Iron Age. The logical question posed to Afrocentrists is though: how was it possible for Africans to pass from Stone Age culture directly to Iron Age culture unless the technology arrived from somewhere else? Did they learn it from the results of forest fire, otherwise? Did they learn iron metallurgy from pottery firing?

According to David Phillipson (2005), it was unlikely for Africans to pass directly from stone to iron production without previous experience of metal working tradition that requires attaining a higher temperature impossible for ancient Africans to produce (see Childs and Herbert, 2005:278). This implies that Africans had to receive the technology of iron production from an area it has developed it earlier. 'The diffusionist model...' argue Childs and Herbert (2005: 278) 'was no doubt buttressed in some quarters by older assumptions about African backwardness..., but it also seemed to have the weight of logic on its side.'

However, despite 'the weight of logic on its side', the outside origin hypothesis of iron technology through pyrotechnology of copper making did not take into account the presence of other mechanisms for transition. For instance, these include the possibility of transition from pottery kilns to smelting furnaces (see Mcosh 1979; Todd and Charles 1978a). The challenging question is, 'could a Neolithic society with no experience beyond the firing of pottery in open bonfires (kilns were not used in pottery-making outside the Nile Valley before 1000 b.c. and rarely thereafter), skip directly to iron working without passing through an age of copper or copper alloys?' (Childs and Herbert 2005: 279-280). It is long noted that ancient potters used iron oxides to decorate their pots as well as their body. Mcosh (1979) articulates that it is possible to produce a heat between 700ºc and 800ºc to produce wrought iron in a small hole in the ground. Hence, it is argued that 'a sufficiently reducing atmosphere combined with the high temperature of the kiln could have reduced iron oxides…' (Todd and Charles 1978a:37; Tylecote 1980) and this procedure could have led to the discovery of metallic iron (cf. Chapter VI).

On the other hand, many scholars have contended that if Africans have moved from the use of crudely manufactured stone technology to a more refined and varied forms in different parts of the continent, why could not they do similarly with the production and utilization of metals? They went on further to ask whether 'decadence' (see Chirikure 2005) could be an explanation for the later prehistory of Africa. The other challenge, as I have already pointed out, for the proponents of both external and the Egyptian origin of iron technology in Africa is the discovery of early dates in different parts of sub-Saharan Africa than the region in the north (see Mapunda 1997; Schmidt 1997; Schmidt and Avery 1978, 1983; Schmidt and Childs 1985). A challenging instance can be drawn from the second millennium B.C date for the Termite hills of eastern Niger (see Childs and Herbert 2005:278 on various views regarding the interpretation of the data). There is particularly no site (e.g. in Egypt) dated earlier than many sites in different parts of Africa (see Mapunda 1997 for the summary of the dates and this will be discussed below in this chapter; see also Killick 2009). Accordingly, Africans in different corners of the continent might have responded to their needsaccording to local environmental requirements (e.g. Mapunda 2000; Schmidt 1997, 2001; Schmidt and Avery 1983; Schmidt and Childs 1985; Schmidt and Mapunda 1997). This buttresses the argument in favor of multiple origins of African metallurgy.

In contrast, early dates (e.g. in central African and sites in the west) faced rebuttal from different scholars on the basis of lack of precision of C14 dating and the problem of *old wood* use for charcoal (see Childs and Herbert 2005; Holl 2008; Killick 1999, 2009; Miller and Van der Merwe 1994; Van der Merwe 1980). However, the argument favoring multiple origins of iron technology has been established firmly on many other grounds than just dates. For instance, according to many scholars, technological variations explain the presence of *long traditions of experimentation, innovation, modification and adaptation of the technology to specific local needs* (see Burka 2006, 2008; Miller 1997; Schmidt 1997; Schmidt and Avery 1978, 1983; Schmidt and Childs 1985; Schmidt and Mapunda 1997; Van der Merwe and Avery 1987). The variations can be found in areas such as various forms of furnace designs, the tuyères, the bellows, and the different forms of cultural manifestations of the production activities (see Burka 2006, 2008 on technological variability among close lineages).

The core emphasis of this dissertation is that its attempt to underline the importance of recognizing various competing views without dipping into one of the extreme sides. This is mainly because the existence of such diversity and the attempt to investigate further on the reasons behind their variability offers an opportunity for developing an alternative model of explanation. My attempt to develop an alternative model of explanation is derived from evidence in two localities of Wollega situated hundreds of kilometers apart but inhabited by metalworkers from close lineages (see Burka 2006, 2008, 2009; cf. *chapters II and chapter V*). The Wollega metallurgists have developed different technological traditions of iron production. In the mean time, I am aware of the challenges that are still undergoing regarding the chronology of African archaeometallurgy (see Killick 2009). By no means, the ethnoarchaeometallurgical study of iron metallurgy in Wollega is about the chronology of its trajectory across time. The question of situating the Wollega material in time trajectory should wait for a comprehensive and concerted archaeological investigation of the wider region in western Ethiopia and laboratory analysis of the finds. I have refrained also from suggesting relative[2] dates of metallurgy areas based on oral traditions and genealogy as well as ethno-history of the region.

Hence, the ethnoarchaeology of iron metallurgy in Wollega is primarily about the technology and the people rather than entanglement with the question of origin. This needs considerably more work in the future. The study, therefore, seeks to understand how the society in Wollega utilized the technology of iron metallurgy and what the technology became for the people after it appeared as a cultural trait. In view of regional population and cultural dynamism in the region, it is not about who brought the technology to and how old it is (cf. chapter VII). At a wider level, the following tracts from a detailed discussion of the challenges of origin of African metallurgy by Childs and Herbert (2005:281; See also Miller 1997, 2001, 2002; Miller and van der Merwe 1994; Killick 1999) best summarize the state of the art in the field and future trends:

> ....If anonymous African metalworkers did indeed master the art of smelting iron ores without previous experimentation with pyrotechnology, archaeology is unlikely to unravel the mystery of how they did it without considerably more work. This must involve more fieldwork in the little-explored regions of the continent, careful cross-checking of dates by several methodologies, and use of archaeometallurgical lab techniques to understand technical achievements...

Nevertheless, in the following sections, I concentrate on the following points: First, I situate Ethiopia geographically and culturally within the context of Northeast Africa. Second, I discuss different explanations offered by various scholars regarding the introduction of metallurgy in the region and evaluate their strengths and weaknesses. Thirdly, I forward an alternative model favoring a moderate stand in the discourse for explicating the presence of complex cultural developments in Ethiopia and the Horn since antiquity. Finally, I

---

2 The Oromo have rich tradition of allowing us look backwards for a long period. According to tradition from the highland region of Horro in northeast Wollega, a religio-political center at Bulluq is credited to the father of Horro and Guduru by the name of Jaawwii. The local genealogy (including those presently over 40 years old) put the succession between 49 years ago and the time of Jaawwii to 14 Lubaas. Lubaa is a period of 40-48 years in Oromo traditional Gadaa system (see Leggesse 1973; Hassen 1990; Tesema 1980; Gidada 1984). If we take the minimum value for the Lubaa (40 years), Bulluq was founded over 500 years ago i.e. 40x14=560 years. If put this in chronology, c.1973-560=c.1413 AD. There is another method, I could add to strengthen my relative dates of the area. According to local history documented by CHernet Waqweyya who interviewed the last Abba Gadaa (Galataa Abba Namaa), it was learnt that he was the 68th cycle when the last celebration was conducted in 1973, a year before the Imperial regime of Haile Sellasie was overthrown, and the Derg froze such celebrations. Since it is known that one cycle equals eight years, the dating is very simple i.e. 68x8=544. In other words, according to this, the center was founded around 1429. Since the site is both ritual and secular and since the ritual heads and the smiths are of similar roots (see Burka 2009), iron production in Abee-Dongoro, which in the later times has served wider areas in the region might have originated not later than the foundation of Bulluq.

concentrate on the following points: the research gap in Ethiopian Iron Age archaeology, major and minor objectives being identified for meeting the stated problems. In order to deal with multifaceted problems connected with a history of iron metallurgy in Ethiopia in general and Wollega in particular and for attaining the selected goals, I will structure the method(s) and techniques I have adopted.

1.2. NORTHEAST AFRICA

The interaction between Northeast Africa and its immediate neighbors such as the Near Eastern peoples has been understood in simplistic terms (see Fattovich 1996, 1997a, and 1997b). First, it simply puts the connection of northeast African regions with the so-called the 'Fertile Crescent' of the Mesopotamian region through Syria and Israel to Egypt (Arkell 1966, Fattovich1997a). Secondly, it ascribes the appearance of state-like society in northern Ethiopia and the Horn in the first millennium BC to the result of such interactions and the 'colonization of immigrants from south Arabia' (see Fattovich 1997a, 1997b; Phillipson 1990, 2009; Schmidt 2009; Sellassie 1972; Van Beek 1967). According to Fattovich (1997a), such presupposition of an historical phenomenon is too simplistic to explain the complexities and the history of dynamic relationships in the wider region. Once more, the aim seems to situate Egypt in the middle between the Near East and the rest of African region.

On regards of technological transfer and Egyptian Pharaonic conservatism towards adopting new ideas (see Edwards 1996; Trigger 1969: 32-33), Egypt might have deterred the spread of the knowledge to the rest of Africa (see Pliener 1980; 2000) than serving as a channel of diffusion. This assumption can be regarded to challenge the *Egypt-as-facilitator* (e.g. Arkell 1968) approach.

However, my current attempt constitutes exposing the limitations of both antagonistic sides- both nationalist Afro-centrists and colonialist diffusionists (see Chirikure 2005; Holl 2008; Shepherd 2002; Trigger 2003). It attempts to underscore the need to provide an alternative means of explanation towards a more comprehensive conceptualization of the regional culture-history based on sources such as archaeological studies on both sides of the Red Sea. In other words, my discussion is short of influenced by specific developments of each region of the northeast. In contrast, I argue that both of the inside and the outside antagonistic standpoints are not exclusive to one another. From geographical and cultural standpoints, in the Northeast African region (the Ethiopian region and the Horn in particular), it is plausible to argue that no single approach suffices to address the regional complex past.

In the 1990s, Fattovich (1996) came up with what he called the '*Afro-Arabian circuit*' model, in his attempts to make sense of the complexities of developments that took place on both sides of the Red Sea since the first part of the Holocene. The model is rooted in archaeological research he has been carrying out since the 1970s. The results of the research led him to argue that the cultural remains such as pottery in 'Pre-Aksumite Culture' indicate an African component. He, however, did not rule out the presence of traits from across the Red Sea or the Nile valley on the African side. The hypothesis contrasted with the dominant view at the time, which attributed any cultural changes in northern Ethiopia to South Arabian colonization (for details see Fattovich, 1978; Selassie 1972; Phillipson 1990; Van Beek 1967).

In his comparison of 'Pre-Aksumite' pottery with the neighboring regions, he concluded that they exhibited 'few similarities with vessels from South Arabia, many affinities with the Sudanese pottery and no connection with the East African ware' (Fattovich 1978:25). Fattovich's viewpoint was intuitive for the period and in my opinion a basis for the general understanding in the 1990s that Afro-Arabian relations were more complex (should be addressed in a multivariate one) than previously put in a simplistic one-way influence from the South Arabian side. I shall summarize below the essence of the discussion revolving around the 'Afro-Arabian circuit' centering on the circulation of various products (see Figure1) across time and space. For now, I will concentrate on discussions regarding the origin of iron metallurgy in Ethiopia and the Horn.

To date, there is no substantial evidence against the established view that the earliest evidence for ironworking (in particular) is found in the Near East dating to the second millennium BC (see Haaland 2004a; Muly1980; Muhly 1980; Pleiner 1980, 2000; Todd 1985; Van der Merwe 1980; Waldbaum, 1980). Separate references of relatively early dates were made for different sites in Africa. However, there is no conclusive evidence as well as consensus among scholars particularly on the early dates (see Alpern 2005; Holl 2008; Killick 1999, 2004, 2009; Schmidt 1997).

Some of these dates include Naukratis in the Nile delta dating to the sixth century BC (see Arkell 1966), Taruga and Nsukka in Nigeria around the sixth century BC (see Okafor 1993; Tylecote 1975), and Buhaya in Tanzania dating between 500 and 200 BC (see Schmidt and Childs 1985). Aksum in the Ethiopian highlands offers an early date of about ninth century BC (see Phillipson 1977). Nevertheless, others argue in favor of fifth to the third centuries BC (see de Contensen1981; Todd and Charles 1978a) or to the second and first centuries BC (see Trigger, 1969). On the other hand, Meroe in the Sudan dates to third century BC (see Shinnie 1985) or since the fourth century BC (see Arkell 1966; Mcosh 1979). Various dates for the Ethiopian region indicate lack of consensus mainly due to the absence of serious research addressing metallurgy (*cf. chapter VII*).

One significant excavation in one of the 'Pre-Aksumite' rock shelters in the highlands of Ethiopia (Phillipson 1977) revealed 'unusually' early dates for the presence of significant iron metallurgy in Ethiopia. Unusually because many researchers focus on the appearance of metal tools or metallurgy in the region, either in the mid or second half of the first millennium BC (see Contenson 1981; Todd 1985; Trigger 1969). Phillipson (1977:78) describes about the presence of metallurgical remains at Gobedra rock shelter in that "these indicators of metal-working [slag and tuyères] are restricted to the two highest strata. Stratum I yielded eighty-six fragments of slag and five of tuyères: There were also thirty-six pieces of slag from stratum II a..."

Nevertheless, Phillipson did not provide detailed discussion on the significance of metallurgy in this site. He asserted, however, that the evidence is only indicative of iron metallurgy not other sorts of metal working. In particular, the discussion fell short of distinguishing between smelting and smithing activities. As far as we know, there was no laboratory analysis of the remains discovered at Gobedra rock shelter nor was a single article written. An indirect inference from the stratigraphic dates of the site suggests that iron metallurgy developed in the highlands of ancient Ethiopia as early as the ninth century BC, which is described as 856 b. c. ±53 years (see *Ibid*:58). Phillipson (1977:81) concludes that the "evidence for iron is common only from this final phase, although a few pieces were also recovered from stratum IIa" (see also Chittick 1978:1).

The indirect date obtained from the Gobedra evidence for iron metallurgy, however, failed to fulfill the hopes expressed then because of two factors: First, in the course of time, it was realized that the finger millet was an intrusive element which made the dating of the site unreliable (see Haaland 2004a; Phillipson 1993; Todd 1985). Secondly, there has not been any follow up work on the nature and significance of metalworking in Ethiopia. As a result, we are still in the dark about the state of the archaeological finds of metallurgy remains at Gobedra. One wonders why the 'finger-millet' problem discredited the whole work on the site. Consequently, and because of the general trend in Ethiopian archaeology, the early history of Ethiopian metallurgy has remained in obscurity or rather 'incomplete' as many scholars prefer to call it (see Haaland 1985, 2004a; Todd 1985; Phillipson 1993).

In spite of lack of complete documentation of the history of Ethiopian metallurgy, the question of its origin has significantly attracted different scholars who propose different hypothesis for its origin. In the following discussion, as a precursor to the alternative model I offer, I will illustrate the hypotheses forwarded by different scholars regarding the origin of metallurgy in Ethiopia and the Horn. The remarks on the region's metallurgical past can be encapsulated into four assumptions (see Burka 2006; Mapunda1997:115-117).

The various hypothesis on the means of introduction of iron metallurgy in Ethiopia and the Horn include the Egyptian origin, the Sudanese origin, South Arabian origin , and an indigenous origin hypotheses (see Arkell 1966; Haaland 2004a; Phillipson 1993; Shinnie 1967; Todd and Charles 1978a; Trigger 1969; Wainwright 1942a, 1942b, 1942c, see also Wainwright 1954 in Todd and Charles 1978a). Previously (Burka 2006), I argued against the independent origin hypothesis viewing the Ethiopian region as never closed to its neighbors out of which my current third model has been developed. However, in this discussion I have revisited this hypothesis realizing that it makes an integral part of the alternative approach to the study of the culture-historical past of the region.

1.2A.THE PTOLEMAIC 'HUNTERS' HYPOTHESIS

The earliest evidence of metalworking in northeast Africa is obtained from Egypt (see Arkell 1966; Miller and Van der Merwe 1994; Tylecote 1980; Van der Merwe 1980). According to these sources, in the beginning, Egyptians, like many societies in other parts of the world, might have worked on meteoric iron (Mcosh 1979; Trigger 1969). As far back as the 19th century BC, for instance, the Bedouins of the Sinai peninsula believed that 'a man who succeeded in forging a sword of meteoric iron was invulnerable in battle' (Mcosh 1979:156). (In the bracket one can cite a similar tradition among the Oromo that relates that a farmer who succeeded in obtaining fragments of a thunderbolt to forge a ploughshare from it or put a piece of the material in his granary becomes a rich person).

Geographical proximity to the Near East and population interactions due to this might have enabled Egypt to obtain the knowledge of iron working from its eastern neighbors than the rest of interior parts of the continent (Arkell 1966, 1968). This might have been much easier for Egypt than other sub-Saharan Africa due to its developed copper smelting tradition (Arkell 1966, 1968; Van der Merwe 1980), a pyrotechnological prerequisite for iron working (Pleiner 1980, 2000; Tylecote 1975, 1980). There is ample evidence that Egyptian metal working tradition goes as far back as the fourth millennium BC during the Old Kingdom (see Arkell 1966, 1968; Killick 2009; Shinnie 1985; Trigger 1969). However, according to available sources, iron tools only appeared during the New Kingdom, in the second millennium BC (see Edwards 1996; Miller and van der Merwe 1994). The first tangible evidence regarding the use of iron is obtained from burial goods of Tutankhamun dating to 1340 BC (Haaland 1985, 2004b; Mcosh 1979). The burial contents include an iron dagger and miniature iron. The gold tinted handle dagger is a foreign origin, according to records (see Edwards 1996).

Nevertheless, the fact is that whether Egyptians worked on terrestrial iron before the 15th century BC is not well documented although some attribute this to a date much earlier (see Mcosh 1979). By 1200 BC, Egyptian metallurgists were practicing with a technique of quenching and tempering iron to harden their products (Miller and Van der Merwe 1994; Van der Merwe 1980). The problem is whether the technique was developed along with a locally smelted iron or not. On the other hand, other authors (see Abdu and Gordon 2004), on many grounds, contest the attainment of such techniques by Egyptian metallurgists. They question whether such technology existed in Egypt as early as the 13th century BC. In addition, they related to the absence of certain dates for the object associated with the technology of quenching and tempering.

In general, iron working has been a rare activity before the Assyrian invasion and occupation of Egypt in the Seventh century BC (see Mapunda 1997). Arkell (1966), however, argues that the significance of the Assyrian invasion was not in the introduction of iron for the first time to Egypt but in marking the beginning of the Egyptian Iron Age. Metal working sites in Egypt during this period were located at Daphne, Nebeshah and Naukratis where Assyrian mercenary armies consisting of Greek, Ionian and Carian metalworkers were stationed (see Arkell 1966; Mapunda 1997; Shinnie 1985; Oliver and Fagan 1975; Pleiner 1980, 2000; Trigger 1969). The contention on the Egyptian origin of iron working is, therefore, that even during the Assyrian occupation, the smiths were not Egyptians (Mapunda 1997; Trigger 1969) but outsiders such as Greeks.

On the other hand, analysis of iron artifacts at Naukratis has shown that the types were Egyptian rather than Greek (Abudu and Gordon 2004). This implies that the role of the Greeks (probably also that of Assyrian mercenaries) was to import iron ingots on which Egyptian smiths worked. This last analysis might also mean that Egyptians were importing ingots to supplement their local products. Pyrotechnologically speaking, however, the Egyptians had a long period of experience with metallurgy (Arkell 1966; see also killick 2009). Given their long period of experimentation with fire to produce other metals than iron, the possibility of local transition to the Iron Age is more likely.

It was based on these fragmentary pieces of evidence that the regions of Ethiopia and the Horn were placed as the end-receivers of the knowledge through Ptolemaic 'Hunters' who diffused the technology during their successive hunting expeditions in the regions south of Egypt. The chief proponent of this hypothesis is Wainwright (1942a, 1942b; see also Mapunda 1997). Wainwright (1942a, 1942b, 1942c) discusses the place of iron in Ethiopian societies ranging from its sources to various local terms for the material (see 1942b). He also mentions the role of iron for the Aksumite kings particularly as exchange item for gold in the remote part away from the center of the empire (*Ibid.* 1942a). This was what historians attributed as silent trade (see Sellasie 1972).

According to Wainwright[3], the Ptolemaic hunting expeditions particularly from the time of Ptolemy III (245-222 BC) were agents for transmitting the knowledge to the local societies of the Horn. For him, the Ptolemaic 'Hunters' must have impressed the local community with their advanced hunting technology. He posits that the 'Hunters' have come from countries where the Iron Age had long been established (Ibid: 1942b:85). According to him, it is possible to presume that the Ptolemaic 'Hunters' used tools and weapons made of iron and were pioneers in introducing the knowledge to the place they have visited.

Wainwright's hypothesis of the Egyptian origin of the knowledge is derived from classical literary documents mainly the *Periplus of the Erythrean Sea* (Mapunda 1997; Todd 1985). According to the *Periplus,* iron was imported by the people of Adulis, (an Aksumite port), for making spears, which they used for hunting and fighting (Casson 1998 in Mapunda 1997). However, Wainwright's assertion of Ptolemaic 'Hunters' role in transferring the knowledge of metallurgy could not address the crucial question of whether the hunters were accompanied by smelters and smiths. Alternatively, has the impression of the new technology just led the local population to acquire the knowledge?

Cognizant of the significance of the information in the classical literary sources such as the Periplus, one can still challenge this hypothesis on various grounds. For example, trusting that the author of the Periplus was right when mentioning the import of iron for making weapons by people in the Red Sea coast, it does not rule out the possibility of iron manufacturing before this date. Was it not possible that the inhabitants who imported iron did so partly to supplement their local products? What explanation is possibly given for iron production sites e.g. Gobedra rock shelter (see Phillipson 1977) where the record shows a far earlier date than the period of the Periplus or that of the Ptolemaics for that matter? What explanation do scholars offer for references that there was an 'advanced metal working' industry by the 3rd century BC (see Contenson 1981; Michels 1990; Todd and Charles 1978a) in the 'Pre-Aksumite' sites of Yeha and Houlti before Ptolemy III? Does it not suggest the possibility that when the people of Mosyllum (known today as Ras Hantarah) were not willing to import very much iron, they did already have an alternative local source to meet their demands (see Mapunda 1997)? What explanation does an ethnoarchaeological study of metallurgy in Wollega provide us with the trends regarding import of the same material that was locally produced?

---

3 W.G. Wainwright (in 1942a, b, and c) is very important for providing necessary information on the introduction of metallurgy (iron) into the northeastern Africa and beyond. Despite his diffusionist standpoints, he has provided a good deal of data in series of publications in the journal known as Man.

Ethnoarchaeometallurgical evidence from Wollega points to the fact that the local iron smiths did not drop local production despite the fact that scrap iron was available to them. Conversely, the smiths were importing scrap iron while they continued to conduct local smelting adding up a new division of labor. According to informants from formal former iron smelters, after they decided to stop smelting, the smiths[4] took up the activity to fill-in the expanding needs for metal products. When, therefore, in the 1950s the smiths/smelters of Western and Eastern Wollega began importing raw iron from outside, did not the inhabitants of the region have hundreds of years of experience in iron smelting?

1.2B.THE MEROETIC/KUSHITIC HYPOTHESIS

Meroe has once been described as the 'Birmingham of Africa' (Tylecote 1975 after Sayce 1912). This implied that as the Birmingham of Great Britain had laid down the foundation for European industrial revolution, Meroe was viewed as the core of iron metallurgy from which the knowledge diffused to the rest of Sub-Saharan Africa (see Arkell 1966). Its iron working industry has attracted significant interest since long period of time (Edwards 1996; Haaland 1985; Mcosh 1979; Shinnie 1985; Tylecote 1975). The theory which pointed to Meroe as the source of the technology of metallurgy has arisen out of the presence of slag mounds indicative of large scale iron production (see Abdu and Gordon 2004; Haaland 1985; Haaland and Haaland 2007; Shinnie 1985; Todd and Charles 1978a, 1978b; Arkell 1968). However, the distribution and use of the implements are poorly documented and became common finds as grave goods only in late and post-Meroitic period (Edwards 1996), which could account for the dispersal of the goods through trade (see Haaland and Haaland 2007; Phillipson 1985). Have the Meroites invented iron metallurgy independently? Was it introduced from the Middle East? According to Childs and Herbert (2005:280), copper objects in Upper Nubia are dated to 3000 b.c. and its smelting to 2600 b.c. They assert that although "iron objects and iron-smelting appear much later, there is no reason to believe that either technology was developed locally...."

According to Mapunda (1997:119) scholars who believed that Meroe has served as centre of diffusion of metallurgy to the rest of Sub-Saharan Africa can be categorized into two. The first are those who have suggested that iron technology and other cultural traits found in East Africa among eastern Bantu speakers (e.g. Sutton 1971; Ehret 1973, 1982, and 1997) were adopted from southern and central Sudanic speakers during their contacts with the Bantu spanning over the period of 2500 years ago. The second group of scholars consists of those who challenged this (e.g. Arkell 1966). They contend that population movements and contacts through the swamps of the Sudd was not possible to the realization of the hypothesis for the spread of the knowledge through speakers of central and southern Sudanic language. Therefore, this group proposed an alternative route, which is first westwards from Meroe

4 As in the next successive discussions I would attempt to show, the transformation of natural ore to bloom and the transformation of the latter into useable objects in Wollega have been conducted by people of distinct specialties. In this regard, when informants mention of the 1950s and the period in the aftermath, smelting has been taken over by the smiths, it points out to different occupational identities of the smelters and the smiths.

(e.g. Cole 1954; Posnansky 1966 on the Bantu migration), and then in the direction of south and southeastwards (Cline 1937; Wainwright 1954 in Todd and Charles 1978a).

Arkell (1966:452), however, strongly contended that (Mapunda 1997), whether the spread of the knowledge has been made east of the Sudd along the foot hills of Ethiopia southwards or first westwards to the Lake Chad and from whence southwestwards, the early Iron Age cultures often associated with the Bantu speakers owe their origin to the Meroitic civilization. On the other hand, Wainwrithgt (1954) has previously provided two alternative routes of diffusion from Meroe to the Lake Victoria region (see Todd and Charles 1978a:32): via the region of the White Nile, possibly through contacts with the Nuba of Kordofan or via the Ethiopian highlands.

According to Wainwright, the Didessa tributary of the Nile [in Wollega] would have been within southeastern reach of the Meroitic culture area (see Todd and Charles 1978a). If Wainwright's Didessa[5] valley hypothesis is considered, then the two Wollegan technologies of iron smelting find their respective explanation. Accordingly, whereas the northeast Wollega underground furnace could be attributed to the pre-Roman period (dated to be some 3-4 centuries before our time); the technology in southwestern Wollega with its splendid superstructure and tapping mechanism could appropriately fit with the post 1st century AD Roman type Meroitic smelting technology. In this regard, iron smelting in Wollega could have been taking place over the last 2000 years. In addition, Wollega could have served as a staging ground for diffusion of the knowledge to southwestern Ethiopia along the Gibe-Omo river system (see Todd 1985 for her comparison between Omo valley technologies with the Meroitic one). Accordingly, if itinerant/renegade (Pleiner 1980, 2000) smiths were responsible for the dissemination of the technology, then the Meroitic iron metallurgical knowledge could not have taken many years for its adoption by the societies in Wollega or beyond. Of what we know in Abee-Dongoro (close to Didessa) from genealogy of former smelters of Malkee lineage and bricolage, iron production has been in practice in the last over 500 years.

Even though many scholars have entertained the idea of Meroe as the 'Birmingham of Africa', it has been challenged because of lack of archaeological evidence for its spread from Meroe east, west or southwards (Mapunda 1997:119). Trigger (1969) in particular challenged the assertion, pointing to lack of any Meroitic cultural trait to be demonstrated in West Africa or any trait indicating that it has diffused from Meroe (see Killick 2009). By the late 1970s, Todd and Charles (1978a) have noted that they were not aware of the presence of any archaeological evidence concerning its diffusion west, east, or southwards from Meroe (but see Todd 1985). Then how is Meroitic hypothesis applicable with regard to the origin of Ethiopian metallurgy? How can one reach at any plausible conclusion in the absence of significant archaeological research in the upper Nile regions of the Sudan or western Ethiopian region?

5 The Didessa (in Oromo Dhidheesa) river valley separates Wollega into east and west zones and extends southwestwards to separate Wollega from Ilubabor and Jimma Abba Jifar zones. It is part of the Blue Nile river system through which Nilo-Saharan speakers of the Gumuz intruded to the Wollega hinterland, which is dominantly inhabited by the Oromo speakers of Macha (Macca) branch (see Bartels 1970, 1983; Gidada 2001; Hassen 1990; Knutsen 1967; Hultin 1984; Ta'a 1980; Triulzi 1981, 1996).

Critics of the Meroitic Hypothesis hold that the assumption that Meroe served as the center of diffusion for iron technology to the rest of Sub-Saharan Africa depends on the large mounds of slag found in place. On the other hand, the claim of regarding Meroe as center of diffusion also affects Ethiopia (Mapunda 1997). Although direct metallurgical evidence that point to connection between Meroe and Ethiopia is absent (but see Phillipson 1993), other cultural traits suggest the presence of interaction between peoples of the region (see Fattovich 1978, 1990, 2009; Michels 1978; 1990; Munro-Hay 1993). From the previous discussion on Egypt, it was related that metalworking in Egypt is attributed to mercenaries such as the Greeks who accompanied Assyrian invasion of the region.

On the other hand, these same Greek mercenaries also took part during the invasion of Napata in 591 BC (Todd and Charles 1978a). According to Todd and Charles (1978a:33), it is about this period that a regular inland trade route established between Adulis [Aksumite port on the Red Sea] and Meroe (see also Fattovich 1996, 1997a, 1997b on the antiquity of interactions and circulation of objects in the sub-region). As far as the Ethiopian region is concerned, by the end of the first century BC, not only a Greek community has been established in the region but Greek has also begun to be utilized on Aksumite coins and inscriptions (see Anfray 1981; Contenson 1981; Sellasie 1972; Van Beek 1967). This, however, lacks strength to explain about records of pre-1st century BC iron metallurgy in different sites in northern Ethiopia (e.g. Contenson 1981; Phillipson 1993; Todd and Charles 1978a). In other words, it could be suggested that prior to establishment of Greek community, an extended period of earlier contact might have existed between the two sides.

On the other hand, it was only in the 4th century AD that reference to iron appeared in the texts from the Ethiopian side in connection with Meroe (see Mapunda 1997; Trigger 1969). In the inscription left by King Ezana of Aksum after his victorious expedition against Meroe that gave it its coup de gràce, he listed iron as among the trophies his army plundered (see Wainwright 1942b). In contrast, as I have discussed in previous section, evidence shows the widespread use of metals in Ethiopia long before Ezana. Minting technology has even advanced much before this time (Munro-Hay 1993). Nonetheless, a number of questions remain to be addressed. For instance, how can we explain the fate of artisans following the collapse of Meroe, the empire that centrally controlled iron smelting (see Haaland 1985; Shinnie 1967)? Was it not possible to suggest that itinerant smiths (renegades) could have spread the skill to different regions even long before the collapse of the empire (see Levin 1974; Todd and Charles 1978a) and more after that? Cannot the same model be applied to the diffusion of the knowledge as in the Near East following the demise of the Hittites in the 12th century BC (see Pliener 1980, 2000; Tylecote 1980; Van der Merwe 1980) that allowed the spread of the knowledge?

1.2C.THE SOUTH ARABIAN HYPOTHESIS

The connection between the regions on both sides of the Red Sea (Ethiopia and the Horn, and Southwest Asia) has provided one of the strongest assumptions regarding the origin of a number of cultural traits including iron metallurgy on the Ethiopian side of the Sea (see Mapunda 1997; Todd 1985; Sellasie 1972; Van Beek 1967). Thus, the South Arabian hypothesis on the source of civilizations, including the knowledge of iron metallurgy, in the Ethiopian region and the Horn, has for long been highly attractive to many researchers (for details see Mapunda 1997; Phillipson 1993; Selassie 1972; Todd 1985; Todd and Charles 1978a,1978b; Trigger 1969). In the early part of the first millennium BC, strong cultural contact seems to have been established between South Arabia (particularly with the kingdom of Saba) and the Ethiopian and Eritrean highland regions (see Fattovich 1990, 1996, 1997a, 1997b; Munro-Hay 1993). By the seventh century BC, immigrants had brought south Arabian literate culture (Van Beek 1967) that retained many artistic, architectural, and technological features from their homeland (see Anfray 1981, and1990 in Phillipson 2005).

According to the proponents of South Arabian origin of ancient Ethiopian civilization (e.g. Van Beek 1967), the first 'wave' of immigrants who carried the various forms of cultural traits to Ethiopia arrived before 1000 BC (see Selassie 1972; see also Todd 1985; Todd and Charles 1978a). In view of the chronology of dates associated with metalworking (see Durrani 2005), the South Arabian source of the knowledge forms a strong basis of argument (see also Mapunda 1997; Todd and Charles 1978a). This, however, is not consistent with lack of serious archaeological investigation focusing on archaeometallurgy in Ethiopia for the purpose of proving or disproving the assumptions favoring the South Arabian source of the knowledge of iron metallurgy (Burka 2006).

Todd and Charles (1978a:33) state that in the 'Pre-Aksumite' site of Haoulti, excavations have revealed 'abundant gold, iron, and bronze metallurgy, dated to the fifth and fourth centuries BC (see also Contenson 1981). In contrast, on the other side of the Red Sea in South Arabia, similar archaeological studies produced results of sites associated with iron metallurgy in the early first millennium B.C (Todd and Charles 1978a, 1978b).

Todd and Charles (1978a:33) take this point further to propose that 'since iron has been found at Hajar Bin Humeid in South Arabia, in levels dated to the tenth and ninth centuries B.C., these dates could also be consistent with the theory that iron working was introduced to Ethiopia via South Arabia.' In addition, Phillipson (1993) inclines to endorse the South Arabian origin hypothesis of iron metallurgy pointing to the appearance of iron in northern Ethiopian region in contexts associated with a strong South Arabian origin. On the other hand, he argues in favor of a Nile valley origin of the knowledge of copper/bronze metallurgy in Ethiopia prior to the arrival of iron. However, given the presence of pyrotechnological prerequisite for iron metallurgy (e.g. bronze/copper metallurgy) in Ethiopia (unless both arrived simultaneously), one wonders why the indigenous Ethiopians should wait for the arrival of South Arabian immigrants to be 'taught' iron working. Does not such an approach denigrate the innovative capability of certain society? In addition, in both cases, copper/ bronze and iron working knowledge, the source of the knowledge points to areas outside of Ethiopia (the Nile valley and South Arabia). The perspective also puts not only northern Ethiopian indigenous societies at the mercy of outside immigrants but creates a mental schema that favors the region serving the role of conduit for the spread of the knowledge to other parts of the population in the region. The consequence of such mental schema can easily be noticed from the way the culture-history of Ethiopian past is structured-developments in the north are earlier in dates than in central and southern parts of the country. In this regard, arguing in favor of the presence of cultural developments in much antiquity, for instance in Wollega, could result in a jolt to the established mental schema in Ethiopian history. Thus, the question of whether or not iron metallurgy has an independent and multiple /internal origins remains unanswered.

1.2D. THE INDEPENDENT/LOCAL ORIGIN HYPOTHESIS

A few scholars argue in favor of explaining the development of iron technology or other Aksumite cultural traits in Ethiopia from an independent origin position. Anfray (1981:377-378) and Fattovich (1978, 1990, 1996, 1997a, 1997b, 2009) argue in favor of this position (see Mapunda 1997). On the other hand, based on radiocarbon dates in the closest region of northeastern Africa 'the early dates for the appearance of iron in Ethiopia indicate that Knowledge of iron technology spread from Ethiopia to other parts of East Africa' (Todd and Charles 1970a:36-7). However, the way Todd and Charles saw the Ethiopian region of northern Horn was not from independent origin stand point. They have already underscored (see the South Arabian hypothesis section) that Ethiopians must have obtained the knowledge from South Arabia where earlier dates for metallurgy is available. Regarding Ethiopia as a point of spread to other parts of East Africa only gives Ethiopia the role of the conduit.

However, Mapunda (1997) has observed the weakness of the independent origin hypothesis in the context of the Ethiopian region in particular (see also Burka 2006). The geographic position of the Ethiopian region *vis-à-vis* regions east of the Red Sea as well as the Nile valley region is a significant challenge to this hypothesis. In addition, the similar socio-cultural developments of the population on both sides of the Sea need critical examination (Irvine 1978; Munro-Hay 1991; Selassie1972; Van Beek 1967) before endorsing hypotheses such as independent origin that favors separate cultural developments (Burka 2006). Mapunda (1997), as a result, cautions the difficulty of ascribing certain cultural traits to solely one region on one side of the sea and denying the other. In addition, because of population and other dynamics in the region since the early Holocene (see Fattovich 1996, 1997a, 1997b particularly his 'Afro-Arabian circuit' model), it is not plausible to argue in favor of a purely separate development of a given culture.

In line with this argument and the **Afro-Arabian Circuit** espoused by Fattovich (1996, 1997a, 1997b), I saw the significant of adopting the ***Third Model*** for understanding developments in the Ethiopian region and the Horn as well as the rest of the continent. The Third Model is an alternative to the two radical positions (outside vs. inside) in explaining cultural changes in the Ethiopian region and the Horn since antiquity. How is the third model presented? How deep in time and far in space should we go to present this model as a plausible alternative point of departure?

1.3. 'AFRO-ARABIAN CIRCUIT': AN INTEGRATIVE EXPLANTORY MODEL OF CULTURAL CHANGES IN NORHTEAST AFRICA

A concerted academic archaeological inquiry into Ethiopia's ancient past was laid down mainly with the advent of the Deutsche Axum Expedition (DAE) at the turn of the 20th century (see Brandt 1997; Crummey 1990). Since then, scholars have offered a number of hypotheses regarding the origin of ancient civilizations in Ethiopian region and the Horn. Central to these discussions have been Aksumite and 'pre-Aksumite' cultural developments (see Phillipson 1990, 2009). The origin of socio-political and cultural complexity in the form of a kingdom in northern Ethiopia and Eritrea with South Arabian features (Fattovich 1996, 1997b, Phillipson 1985, 1993, 2005) has many implications. Firstly, it has always been regarded as an implantation of an overseas civilization in Ethiopia (e.g. Van Beek 1967; see Phillipson 1990). Secondly, this has been attributed to the migration and settlement of people from South Arabia (or Egyptian artisans) in northern Ethiopia bringing with them different cultural traits (Bent 1893/96; Selassie 1972; Vsn Beek 1967). These changes have taken place since the late prehistoric to the early historic times (see Fattovich 1978, 1990, 1996, 1997a, 1997b, 2009 for entanglements with explication of the interactions and the respective cultural changes).

According to Fattovich (1996, 1997a, 1997b), the pattern of Afro-Arabian circuit connecting the Persian Gulf with the Red Sea region since the early Holocene, first derived from exchange of obsidian tools, was more complicated than previously stated. Fattovich has classified the contacts into five stages ranging between the seventh millennium BC (began with circulation of obsidian tools) and the second millennium AD: Stage I (seventh-fourth millennium BC), Stage II (third /second millennium BC), and Stage III (first millennium BC), Stage IV (first millennium AD) and Stage V (in the second millennium AD).

Nevertheless, due partly to the focus of the study on Iron Age archaeology, in explicating the significance of the model as an alternative model in addressing regional cultural changes, I focus on the period ranging from the second stage in the third millennium BC to the fourth phases in the first millennium AD. Yet, to convey a vivid picture of the contacts and interactions between the Ethiopian region and the rest of the neighboring regions, I believe it is helpful to address internal dynamics primarily or together with the influences from external elements (see for example Levine 1974).

As early as the 1970s, Donald Levine (1974) in his Greater Ethiopia offered a comprehensive explanatory model on the peopling of the Ethiopian region and the Horn. According to him, the backbone of the present inhabitants of the Ethiopian region was formed as early as the third millennium BC constituting (see also Fleming 1976; Todd and Charles 1978a) 'peoples practicing rudimentary forms of agriculture and animal husbandry and speaking three branches of Afro-Asiatic-Semitic, Cushitic, and Omotic' (Levine 1974:28; see also Ehret 1997) languages. He maintained that 'proto-Ethiopians' began to separate as early as the third millennium BC forming five core groups: northern Cushitic, Central Cushitic, Eastern Cushitic, the Omotic speaking and the Semitic-speaking ones. Accordingly, he concludes that, the present distribution of Ethiopian peoples, their cultures and languages should be explained as a function of internal differentiation and diversified interactions with external influences (Ibid: 28-29). However, Levine's model of the peopling of the Ethiopian region has been strongly criticized (e.g. Baxter and Blackhurst 1978; Fleming 1976; D. Todd 1978) for its superficiality.

Despite other shortcomings, Levine rightly argues that Ethiopia has never been entirely isolated from the rest of the world (see also Burka 2006). Ethiopia is a region that constitutes a diversity of landscape characterized by rugged mountains, plateaus as well as inhospitable arid low lands. However, people have crossed and re-crossed rugged mountains, deserts, and rivers since distant prehistoric times (see Mapunda 1997, for discussion on interaction in the Horn and its neighbors). In other words, physical features of the land have never stopped people from moving around. In view of this, it is not plausible to assert that the geographic position of Ethiopia has deterred people from crisscrossing it or occupying it temporarily as well as permanently.

If Levine (1974:30) is right, although environmentally deterministic standpoint, the significance of such peculiar Ethiopian physical feature has only allowed the indigenous peoples from being 'overwhelmed' by the outsiders. Nevertheless, the outside influences were attributed to contacts, interactions, movements, and settlements of peoples forming more thick layers on the existing population. These include the Sudanic peoples from the west, the Oriental Semites, and Mediterranean influences (see Fleming 1976; Levine 1974: 30-33; Todd and Charles 1978a: 34-35) with the rest of linguistic groups indigenous to the region forming the underlying layer.

Now, I argue in favor of allowing a flexible model for understanding what contributed to the shaping of the present culture-history of the Ethiopia region and the Horn. A summary of the 'Afro-Arabian circuit' is found in the following Figure 1. This includes the period between the third Millennium BC and the first Millennium AD. The assumption is that anything that had happened as early as the periods in one of the regions has affected directly or indirectly the rest of the regions in the circuit. Put differently, since the Ethiopian region and the Horn has never been 'closed' to its surrounding regions, the development, or innovation of a certain piece of civilization is likely to be diffused throughout the system as well as a simultaneous development of certain cultural traits across the regions.

In the context of the key issues raised by my dissertation, it is plausible to argue that *metalworking knowledge, if it had first originated in Egypt, the Sudan, Ethiopia or South Arabia, it could have circulated from one to the other like or with the rest of the objects, ideas and the peoples.* Technological variability can be attributed to people's desire and innovative capability to adapt itaccording to respective local realities. The lack of ample evidence about metalworking in the heart of ancient Ethiopian empires such as Aksum should be found in the exclusive focus of the archaeological research on the elite areas (see Phillipson 1990, 1993) but not on the lack of the tradition of metallurgy. The following discussion encapsulated in the table (Figure1) shows the extent to which the regions around the Red Sea have been integrated at least in the last 5000 years.

No.	Period, general Specific	Regions congruent in cultural development	Common features	Specific Features	References
1	3rd Mill. BC	South Arabia and the Horn (source: S. Arabia, Africa e.g. Shoa	Obsidian Exchange Contd.		
2	Late 3rd-mid 2nd mill BC	Egypt, punt, Nubia E. Sudan, L. Basaka, Asa Koma (near) lake Abe, and Yemen	Ceramics from E. Sudan Ethiopia, Djibouti & Yemen (Bronze Age Pottery-2900-1800)	Pottery Decoratio Like Gash Group	Intense Contacts (2500-1500)
3	Late 3rd-Mid-2nd mill And in the 1st Mill	**Gobedra** (Tigrean plateau And **South Arabia**-came For transport-1st half 2nd Millennium BC)	Camel tooth with microlitich,	earlier evidence of pottery, Metal working	Connections between the two
4	2nd millennium	Hamasien, Tihama, Kerma C-group, Gash Group in Kassala	-Ona Group-A (1500-1000/900) -Tihama cultural complex	Polished, : Slipped w Lithics frc Tihama	An African Background For Tihama
5	1500-1200/1100	Sihi Adulis, Nadi Urq', Metera, Yeha, Sabir, Kassala (M.Teglinos) -Southwestern Arabia &	Ona Group-A Representative of Coastal Complex site	Incised rim or Shoulder band Polished axes, Small chipped Stone bull Heads	-on the basis Of two radio Carbon dates From Sihi & Sabir

(continued on next page)

(*continued*)

No.	Period, general Specific	Regions congruent in cultural development	Common features	Specific Features	References
	12th century BC	Eastern Ethiopia, regions Of Harar, Mosna'	-Stone dolmens (upright Stone monuments Supporting lintel stone	-Egypt-Punt Trade collapse S. Arabian Control of Trade with Med. Sea	-cultural & Chronological Relationship Not clear
6	1st Millennium BC	-Kdms arose in S.A. :Saba (earliest c.800-400) Qaraban, Hadramaut, Main & Himyar -Northern Ethiopia: Kdm of Daamat (c. 700-400BC)	-Interaction between The shores of the Red Sea might have caused An Ethiopian Sabaean Stat -The region was Included before the Consolidation of Daamat	-worship sout Arabian gods -inscriptions -epigraphic Evidence of Four kings (S.A.=Mukari -Monumental Architecture Monumental Inscriptions Art, votive Incense altars Offering table Bronze seals, Writing	-superficial -impressive -affecting only the elite -few vessels are Arabian -Main features Are Ona Grou A -similar to the Late prehistori Tools of the Sudan and The Horn

(*continued*)

No.	Period, general Specific	Regions congruent in cultural development	Common features	Specific Features	References
7	4th C BC	Ptolemaic appearance From Egypt to Rhapata (early 1st C AD)	Hellenistic sources Tigrian Plateau isolated -Ptolemaic ships Frequented .Roman trade	Daamat Apparently Collapsed	South Arabian Archaeological Evidence
8	1st Mill AD	Consequences of Roman Trade, connection with Eastern provinces of The Roman Empire (Egypt and Syria)	Aksum Amphoras, Glasses Christianity -Aksumite elite residence -Roman villas in S. Syria Churhces=N. Syrian Churches (round or Square apse of 4th and 6th c AD)	-Aksum Roman mai Partner in Africa -e.g. Kaleb' Tomb Constructio Technique= 6thc monum in Syria	-Egypt and Syrian origin

Figure 1 Summary of the 'Afro-Arabian circuit' based on archaeological evidence for the antiquity of contacts, interactions, and circulation of objects on both sides of the Red Sea

1.4. DEFINING THE PROBLEM

Most archaeological studies on Ethiopia focus on elite architectural and monumental sites (Burka 2006; Finneran 2007; Phillipson 1990, 1993, 2009). This is evident from over a century-old research known as (pre)Aksumite archaeology (see the works of Anfray, Fattovich, Phillipson, Contenson, Chittick, Sellasie, Munro Hay; Michels on the Tigrean plateau cited elsewhere in this thesis). The focus on elite exotic goods and the outside origin of Ethiopian civilizations (the legend of the Queen of Sheba and King Solomon in particular) have mostly ignored wider perspectives on the regional past (e.g. Chittick 1974; Contenson 1981; Munro Hay 1991; Sellasie 1972). This elite based focus can be noticed from lack of a single article devoted to the significance of metallurgy in the socio-economic and cultural development of ancient Aksumites where century old research concentrated (see Burka 2006).

In line with the above discussion, it is significant to uphold the perspective that the purely ***exogenous approach*** not only neglected Ethiopia's geographic position (see Burka 2006; Fattovich 1996, 1997 a, 1997b; Levine 1974; Mapunda 1997; Mitchell 2005; Selassie 1972; Todd 1985; Todd and Charles 1978a) but has also sidelined views that favored innovative capacities of the indigenous people. In addition, the approach has created a vague picture of the ways of life of the people as though the exotic imported goods were representative of the culture of the whole society. The other problem of Ethiopian archaeological research since its formal inception in 1906 has been its northern Ethiopian regional bias (Burka 2006; Finneran 2007) as well pushing the archaeology of metallurgy into the backwaters (see also Burka 2008, 2009, 2010). In other words, the archaeology of Ethiopia suffers from spatial and topical imbalance in its research focus.

Thus, the southern and western parts of the Ethiopian region, including the current study areas in Wollega, constitute one of the peripheries not only in the archaeology of metallurgy but also in other sub-fields of the discipline. *The archaeology of iron production in Wollega can therefore be regarded as a drop of ink into the ocean of Ethiopian archaeology. This is simply because the western Ethiopian region is entirely underrepresented in the country's archaeological study.* Consequently, the study of Iron Age archaeology in Wollega suffers from such serious limitations.

Lack of previous archaeological research on the area demands focus on the ethnographic data regarding the practice of iron metallurgy and the means of its distribution. Nevertheless, lack of previous archaeological research in Wollega implies that it is difficult to situate the finds in a temporal pattern. Critics can argue that in the absence of any previous work, it is difficult to reconstruct a long-term perspective for iron working tradition in Wollega. However, I can still maintain that the absence of previous archaeological work should not discourage the beginning of work from scratch as all archaeologies primarily began from scratch.

The present dissertation has primarily originated from scanty information obtained from literary works. The literary sources which enumerate various items of trade that originated from Wollega and destined for the Sudan and across the Blue Nile in Ethiopia (see Ahmad 1980; Triulzi 1973, 1975, and 1981) only make passing reference to iron among the goods. On the other hand, there are reports of smelting activities in different parts of central and southwestern Ethiopia before the activity stopped in the second half of the twentieth century (e.g. Cline 1937; Haberland 1959; Hamrla 1964/ 1966; Huntingford 1955; Lewis 1965). In the following section, I will highlight on the existing patchy ethnographic evidence on iron metallurgy in central and southwestern Ethiopia. The discussion will be further explored in connection with the ethnoarchaeometallurgical case studies from Wollega (cf. chapter V). This would help in identifying the remaining gaps that the current research endeavor aspires to fill. Because of lack of research on the neighboring regions north of the Blue Nile (e.g. Gojjam), however, this study lacks any regional representation to make a wider comparison.

1.5. PREVIOUS WORKS: THEIR STRENGTHS AND GAPS IN RESEARCH

According to Cline (1937), the Oromo of Shoa, in central Ethiopia, constructed their furnace from pounded or burnt clay, 2 meters high with two openings at the base. The large opening is designed for feeding. He further notes that the furnace was built at the mines and demolished after each smelt for extracting the bloom. Regarding the charging mechanism, Cline mentions that they rarely kindled with bag bellows; otherwise no bellows (which in other words means the furnace was designed to operate in a chimney style). In view of geographical orientation of Shoa, Cline rules out the possible source of the technology from east and central Africa, where such mechanisms dominate, and favored its origin from Arabian side. He also adds that the smelting takes place between 4-5 hours, to extract a spongy bloom beaten free of dirt and slag (Cline 1937:47). A number of issues remain to be addressed from Cline's description of the furnace of Shoa Oromo that I further expand on under chapter V.

Except for Haberland (1959, see also Haaland 2004a, b; Todd 1985; Todd and Charles 1978a; Todd D. 1978), who observed and recorded smelting activities in the Omo valley in the 1950s, it is unclear if other writers like Cline (1937), Huntingford (1955) and Lewis (1965) personally observed the process or obtained the data from secondary sources. These sporadic reports on indigenous smelting in south and central Ethiopia had impacts on future research in Wollega and the rest of the sub-region. First, papers emerged highlighting the significance of iron items in the long distance trade from Wollega (e.g. Ahmad 1980; Triulzi 1973, 1975, and 1981; Waqweyya 1988). In addition, the evidence for local exploitation of iron ore has been reported from different sites in southwest Wollega (e.g. Hamrla 1964/1966).

However, except for Waqweyya (1988) from northeast Wollega close to the iron production sites and Hamrla (1964/1966) for the Aira-Gulliso area, the remaining writers were too general regarding the location of the metallurgy centers. The second consequence of such scanty information was the beginning of ethnographic documentation as of the mid-1970s (see Burka 2006, 2008; Haaland 2004a; Haaland *et al* 2000, 2004a, 2004b; Todd 1985; Todd and Charles 1978a, 1978b) involving two distinct geographic and ethno-linguistic areas.

On geographic and ethno-linguistic grounds, the ethnographic observation and record of iron metallurgy in Ethiopian can be categorized into two. Geographically, both areas are located south of the Nile; but they are far separated and located in different watersheds: southwards to the Omo valley feeding to Kenya and westwards to the Baro and Blue Nile rivers into the Sudan. The first category goes to the Omo river system that flows southwards to Lake Turkana in northern Kenya (Haaland *et al* 2004a; D. Todd, 1977, 1978; J. Todd 1985; Todd and Charles 1978a). Ethnographically also, this group belongs to an Afro-Asiatic sub-group of the speakers of the Omotic languages, confined within Ethiopian region of southwestern part of the country (Ehret 1997; Fleming 1976; Haberland 1959; Huntingford 1955; Levine 1974).

The second category of ethnographically known iron production sites belongs to the Wollega region inhabited by the Oromo language-speaking group that extends to the Sudan in the west and separates from the north by the Blue Nile. Ethnographically speaking, both smelters and smiths speak only Oromo language, a Cushitic sub-group of the Afro-Asiatic super family (see Burka 2006; Fleming 1976; Levine 1974).

Such differences of both regions evidenced the existence of drastically different technology of iron production, other non-physical components, and the power that maintained the redistribution system (e.g., regarding Omoitc smelters/smiths see Haberland 1959; Haaland 2004a, b; Haaland et al 2004a; Todd 1985). With respect to technological variability, one also notices such occurrence within the Oromo speaking producers. Two ethnographic observations and documentation of smelting activities in two different parts of Wollega in 2005 and in 2007 (Burka 2006, 2008) were the bases for the current research works for the dissertation. Nevertheless, a number of issues are still requiring to be addressed. Out of preliminary research attempts, the following research questions have been developed and outlined.

1.6. RESEARCH QUESTIONS

The archaeology of iron production in Ethiopia, through ethnoarchaeometallurgical case studies in Wollega, attempts to deal with the following issues: What can the utilization of different technological variables in iron production tell us about the origin of the knowledge? Why did smelters from two districts in Wollega, belonging to closely related lineages, adopt different technologies? Can the model of **population movement** in western Ethiopia (e.g. in the 16th century) be used as the basis of explication for the spread of the technological knowhow? What changes occurred and discernible in iron production activities across Wollega? What aspects of the production activities were maintained and/or changed? What technological/cultural significance does have the use of long tuyères and furnaces with or without superstructure by smelters? How is furnace body treatment integrated into iron smelting technology and interpreted by the owners? How are traditions surrounding iron production reflected in other aspects of society's ways of life (farming, harvesting, ritual engagements, and human procreation)? How is metal viewed in various contexts (e.g. secular and ritual)? How is the social status of smelters/smiths *vis-à-vis* the rest of the population explained? What was the degree of commensality between the artisans and the mainstream society in Wollega? What was the extent of right to own property by the artisans? What is the myth of origin of the group(s) best associated with iron metallurgy? How did both smiths and members of the mainstream explain the historical root of the current marginalization of the *tumtuu* particularly the smiths?

1.7. RESEARCH GOAL

As I mentioned above, the archaeology of iron production in Ethiopia is in its formative years. In other words, a lot is expected to be done for fairly addressing the issue of Iron Age archaeology in Ethiopian region in general and western Ethiopia in particular. It is not enough to note that the Wollega region of western Ethiopia has been a significant source of locally smelted iron to intra and interregional market. This dissertation can be regarded as the first attempt to bring into light the significance of the topic under discussion. As a result, it involves a difficult attempt of putting together a number of fragmentary data in this field of study. Consequently, the following are major and minor research outlines for attaining the goals:

- investigating furnace designs and the different types of tuyères used in both Wollega smelting areas and discuss the possible reasons for the variability between closely related lineages as well as comparing these with data from outside Wollega;
- Evaluating change and continuity in the finds of former metallurgy sites with the ethnoarchaeometallurgically recovered data in terms of house construction, ore utilization, furnace design and decoration, tuyères and charcoal making;
- Developing a technological perspective on the significance of tall furnace superstructure of Aira-Gulliso and the underground furnace of Abee-Dongoroo as well as on the significance of long clay tuyères ;
- Assessing how iron smelting processes and the objects involved in the iron production reflect the culture and thought of the society in the region at large. This is executed through investigation of the symbols, the terms, the taboos, the songs, and the legends about the origin of the group and the technology;
- Investigating the physical location of iron production sites *vis-à-vis* the residential areas and the role of the artisans in various ways of life of the society. Location of metallurgy sites *vis-à-vis* the habitation sites has been regarded as a means of establishing whether thesociety was concerned about the 'pure' and 'impure' setting for iron production;
- Examining the myth of common origin of the *tumtuu* group and traditional ritual (religious) heads in the Oromo of Wollega;

1.8. METHODOLOGICAL FRAMEWORK

Two issues demand synthesis in the search for recognized and plausible approach appropriate for ethnoarchaeological study of iron production. First, it is important to recognize the complexity of the problems arising from investigation of the ethnoarchaeology of iron metallurgy. In this case, I share the concern of Braudel (1980:15 in Schmidt 1997:23) who once observed that research problems are all complex 'growing in scope and depth' and, therefore, cannot be limited to 'a single framework'. The second one is regarding the challenge to address what meanings the technologies constituted and conveyed through their attributes. In this regard, I make note of what Martinon-Torres (2002:30) states that 'technological activity is constituted within an influent social and historical context and... that these fields cannot be analyzed in isolation.' Technology is not an isolated phenomenon from the intricacies in a society (*cf. chapter III*).

In light of the complexity of any research problem, the intertwining of technology with other social and historical matters and above all the fact that archaeological inquiry in Wollega is from scratch, I have recognized the significance of employing a variety of research approaches. The archaeology of iron production in Wollega, therefore, requires a combination of various sources and approaches ranging from ethnographic and ethno-historical evidence to analysis of archaeological data from old smelting sites. However, the use of laboratory tools for macroscopic and microscopic analysis of the finds is farfetched. As a result, an ethnoarchaeometallurgical approach well informed by the *chaîne opératoire* framework for the study of technology (*cf. chapter III*) becomes central to the dissertation. In the following section, I present first ethnoarchaeology and its limitations according to its critics. Then, I discuss how and why ethnoarchaeology is an essential methodological approach for the study of traditional metallurgy-constituting technological and socio-cultural aspects- such as the Wollega ones. For this purpose, I begin by offering some definitions to the concept.

Ethnoarchaeology is the blend of two terms: ethnography and archaeology (see Agorsah 1990; Barndon 2004; Haaland 1988; Schiffer 1978; David and Kramer 2001; Weedman 2004). As an approach, it is a means of using contemporary material culture in order to develop a general or specific interpretative model for material remains of the past. As explanatory instrument, it serves the interest of an archaeologist where s/he faces descriptive and explanatory difficulties of past ways of life from material remains. Thus, ethnoarchaeology enables to observe how extant behavior through material culture is used to reflect on the finds of archaeological investigation and provide a deeper insight and wider perspective towards the surveyed/excavated materials. In other words, the reason for which archaeologists employ an ethnoarchaeological approach as a theoretical and methodological framework is to provide a model for explaining ancient human behavior in the archaeological record (Agorsah 1990; Barndon 2004; Kramer 1985; Longacre 1991; Schiffer 1978; Weedman 2004) as well as studying specific human behavior.

Consequently ethnoarchaeology can be defined as 'the study of [ethnographic]material culture in systematic context for the purpose of acquiring information both specific and general that will be useful in archaeological investigation' (Schiffer 1978; see also David and Kramer 2001; Stanislawski 1978; Staski and Sutro 1991; Longcare 1991; Khan 1994; Kramer 1985, 1996; Weedman 2004; Burka 2006, 2008). What makes ethnoarchaeological investigation peculiar to studies conducted by ethnographers is whereas the research agenda set by archaeologists is to explicate prehistoric artifacts (see Haaland 1988; Kramer 1985; Gould 1978) the focus of ethnographers is different (see London 2000). In other words, ethnoarchaeology is a mechanism of searching for answers, by archaeologists, to the problems connected to interpreting material remains from the past.

Thus, through the study of contemporary material culture the knowledge about past behaviors transmitted in the medium of the objects are investigated. According to Schiffer (1978), ethnoarchaeology enables to create models about the past and that the models are becoming significant when they are situated within theoretical framework of the research topic. Ethnoarchaeology is, as a result, an approach(s) for materializing our theoretical perspectives during the investigation of human learned behavior in the medium of material culture.

Arguably, the study of African metallurgy has been approached by separately treating the technological (physical) aspect from the socio-cultural (non-physical) ones that surround it. In contrast, one can posit that what can be regarded mundane (e.g., ritual) aspects cannot readily be separated from mechanics of production activities (see Barndon 2004; Mapunda 2000; Miller and Avery 1987; Childs and Killick 1993). These categorizations include the physico-chemical and the socio-cultural aspects of metallurgy (see Barndon 2004; David and Kramer 2001; Haaland 2004a; Haaland *et al* 2002; Haaland *et al* 2004b; Burka 2006, 2008). From African traditional metalworkers' standpoint, the separation of the physical from the non-physical aspect is inconceivable (see Burka 2006, 2008; Miller and Avery 1987). This assertion about African traditional metallurgists was made evident during reconstruction of smelting activities in Wollega. This partly explains why ethnoarchaeology is an indispensable approach for the study of metallurgy in Wollega, Ethiopia and elsewhere in Africa (sees Agorsah 1990; Atherton 1983; Clark 1981).

1.8.1.THE QUEST FOR ETHNOARCHAEOMETALLURGY IN WOLLEGA

The *raison d'être* for ethnoarchaeometallurgical approach in Wollega is firmly anchored in many research opportunities and challenges. In the first instance, to the late 1960s, a traditional metallurgical activity has continued to operate in many parts of the region. The question of for how long it operated in the region is unattended (see Footnote 2) except that it is ascribed to distant ancestors. Available literary and oral sources indicate that the product of iron production, from different parts of the region, has widely circulated over extended areas crossing local and international borders (see Burka 2006, 2008; Ahmad 1980; 1999; Hassen 1990; Triulzi, 1973, 1975, and 1981; Waqweyya 1988) at least in the last three centuries.

The significance of this premise relating to its continuity is that the observation of behaviors surrounding technological and socio-cultural aspects of indigenous iron metallurgy can be documented through **reenactment/ reconstruction** of smelting activity. The excellent opportunity is that old smelters who conducted the activity as well as other persons who observed the activity before it stopped in the late 1950s and early 1960s are still accessible to provide first hand information and demonstrate their skills.

This, on the other hand, underlies that there is urgency for documenting the indigenous knowledge of iron metallurgy. One of the reasons for the need to document as soon as possible is, connected to, the dwindling age of repositories of the knowledge. Ethnoarchaeometallurgy is significant to revitalize indigenous knowledge. Only few old smelters have survived. Two old smelters from northeast and southwest Wollega, who have provided moral and intellectual support for the 2005 and 2007 (Burka 2006, 2008) re-enactment and reconstruction projects, are now dead (in 2007 and 2008). This project is mainly important to document the knowledge and skill of indigenous iron production before it is buried with the old smelters and serve as an interpretative model for future archaeological research on metallurgical sites. I do not intend, however, to claim that the knowledge and skill of the technology remained static throughout the ages or that the old smelters remember everything about the technology and its socio-cultural aspects.

I am in line with the logic that ethnoarchaeological approach must distance itself from treating 'traditions' as pristine islands unaffected by the modern world including capitalism, globalization, or cultural innovation. In line with Atherton (1983:76), I argue, however, that the details of manufacture, use, disposal, and life spans of material culture can be lost as time passes. In spite of this, it is still possible in most traditional societies [as in Wollega] to recover an enormous amount of data pertinent to traditional material culture (see also Agorsah 1990; Burka 2006, 2008, 2009; Clark 1981; Phillipson 1985; Schmidt 1997).

The other *raison d'être* pertinent to adopting ethnoarchaeological approach is, like many of the African societies, the core of the old generation society of Wollega is non-literate. One could have assumed that the fact that traditional metallurgy continued until 40-50 years ago means that the technology and its socio-cultural aspects might have been left in written records. On the contrary, the characteristics of a non-literate society is not only that few is repository of society's knowledge (like the old smelters who volunteered to coordinate the reenactment/reconstruction activities for this research) but that this knowledge is passed in the form of oral histories and demonstrations of the activities. Therefore, I was convinced that through ethnoarchaeological investigation of contemporary Oromo materials surrounding the re-enacted/reconstructed iron production activities, it was possible to establish an interpretative model on how the society during prehistoric times had responded to the requirements and constraints of its environment.

Moreover, the choices they made, what constrained them in the course of activities and the various innovative mechanisms they devised in response to various situations can possibly be reconstructed through the *chaîne opératoire* and ethnoarchaeological approaches. The presence of technological and cultural variability in the society, for instance, can be revealed by examination of activities at each stage of the *chaîne opératoire* of production, use, and re-use. Moreover, it is difficult to imagine, by an archaeologist (without ethnographic evidence), that such a big technological rift exists in iron metallurgy between closely related lineages in Wollega. An ethnoarchaeometallurgical study provides a better explanation for an underground pit with remains of iron production. These uncertainties can include whether it is slag pit or a furnace or of some ritual significance other than technological.

Finally, there remains to be the issue of cultural dynamism and innovation. One factor that might have caused old smelters, according to informants, to drop the smelting practice is attributed to the appearance of 'cheap' scrap iron for the smiths. The introduction of 'cheap' scrap iron (from car leafs of the Italian period) has said to have disengaged the traditional interdependence among stakeholders of iron production activities particularly between smelters and smiths. The traditional reciprocity between smelters and smiths has been violated during this period.

Furthermore, the introduction of Protestant Churches, accompanied with modern schools, has also broadened group's field of specialization. This is particularly the case in southwest Wollega. The children of former smelters and smiths could forge different occupational identities that were made possible through achievements in free education. Since such dynamics reoriented ones destinations, the local activities were dropped, and the young generation were far from being instructed or physically present to observe the activity or participate in it to keep the tradition and pass them for posterity. Therefore, ethnoarchaeological project, involving re-enactment/reconstruction of former activities, has not only the role of documenting the knowledge on the edge of disappearance or helping archaeological interpretation but also creating/forging a link between generations by refreshing memories.

One of the critics charged against ethnoarchaeology is that it would only inform synchronic meaning and lack archaeological significance in the distant past or the period of historical antecedents. In addition, though erroneously, ethnoarchaeology is regarded as an attempt by archeologists to extrapolate data obtained from ethnographic present for spatial and temporal comparisons (see Agorsah 1990; Atherton 1983; David and Kramer 2001; Haaland 1988). They also doubt that ethnoarchaeology is replicating the same course of activities by ethnographers.

This is, however, not the case in that ethnographers were not concerned with material culture and devote their works to religions, politics, and sex (see London 2000). Despite an array of criticism against ethnoarchaeological approach, it provides an alternative means of obtaining data in place to use it for law making in the interpretation of prehistoric materials (Agorsah 1990; Clark 1981).

However, if ethnoarchaeological approach were integrated with other methods of data analysis (literary, oral histories, archaeological survey of sites and analysis), the amount of errors we make would be minimized. The integration of an ethnoarchaeological approach with other measurable tools is proven to be securely successful in areas where such research activities were employed (e.g. Barndon 2004 in Tanzania; Chirikure 2005 in Zimbabwe; Miller 2001, 2002 in South Africa; Schmidt 1997 in Tanzania).

Ethnoarchaeometallurgical observations provide a number of activities and associated aspects at a certain point and time. For instance, the meanings of body decorations on the smelting furnaces are attested only from information obtained in the living societies. It is true that changes are inevitable throughout time. Nevertheless, in some archaeological contexts, it is possible to excavate a pit in the ground close to a hearth or a hole in the ground with metallurgical remains at its foundation. What kind of explanations could be drawn from such encounters? Such encounters could be explained securely if ethnographic data is integrated with archaeological evidence. It is through ethnoarchaeological research among extant societies that we could be able to learn that African iron metallurgists employed a variety of furnaces and tuyères in iron production activities. This exposure laid the foundation for strengthening the multiple origin hypotheses regarding the source of the knowledge of iron metallurgy (Burka 2006, 2008; Cline 1937; Haaland 1985, 2004; Haaland *et al* 2002; Haaland *et al* 2000, 2004; Sutton 1985; Schmidt 1997; Schmidt and Avery 1987).

It became clear for me that the community regards non-physical aspects of iron production or which do not seem to make parts of the physical inputs as indispensible ingredients of the activity. My previous preliminary investigation of iron smelting, in eastern and western parts of Wollega, is an impetus for founding an interpretative model. It has provided me with an opportunity to observe many rituals, embodied in iron production, are also reflected in different areas of socio-economic activities among the Oromo (Burka 2006, 2008, 2009). This study has, therefore, expanded my perspective to include other production areas that follow similar procedures like iron metallurgy.

1.9. DATA ACQUISITION TECHNIQUES

As I have already pointed out, the study of iron production is complicated by the need to concurrently addressing the technological and socio-cultural aspects of metallurgy. Unlike in the non-African western world where technology can be viewed just as one of scientific gestures, the traditional African iron metallurgy, from owners' standpoint, is practically impossible to deal with the physical aspect in isolation of the socio-cultural/ritual attributes (Childs and Killick 1993; Killick 2004; Miller and Avery 1987). Even more, the socio-cultural aspects are too imbedded into the physical aspects to be separated and treated on their own or *vice versa* (Burka 2006, 2009). What many might regard trivial does not only constitute the technological categories but production seems to be impossible without its presence (Miller and Avery 1987). This nature of the activity, therefore, demands the adoption of various mechanisms of enquiry techniques to arrive at plausible explanation of the material record in its distinct cultural setting.

In this study, I have made use of different techniques for retrieving the ethnographic data of iron production in its particular ethnographic setting. To attend to the various forms of challenges by the people, I employed various approaches that range from passive observation to active involvement through forwarding different questions at different stages of the *chaîne opératoire*. The following questions were presented to participants of the reconstruction and re-enactment of iron metallurgy and elders at individual and group levels. The same questions presented to various individuals at different times and places were believed to enable to examine the nature of differences or similarities in their responses. Attempt has been made to entertain views from both the mainstream (e.g. smelters) and the rest of the artisans, who due to various reasons were put on the edge of the society.

1. What is the local term for iron smelting? This question enables to primarily distinguish between smelting and smithing; hence the significance of the term.
2. What do you call the clay for furnace and tuyère construction? Are they all the same? Although similar term is used to identify clay in different parts of Wollega, some additional probe was necessary for distinguishing between different types of clay soil for various purposes. This also enables to assess the level of connection between furnace builders and tuyères makers and potters.
3. What is the local term for iron ore? How do you extract it? Who did the extraction/mining? How many places did you have for mining? How did you obtain it in the past? Was it free of charge? Was it for sale? Who owned ore mining sites in the past?
4. Who was preparing the charcoal? What do you call the name of the tree species for charcoal making? Why was this tree species selected? Why were not others? Did you use the same tree for iron smelting and smithing?
5. When did members of your group begin to work on iron metallurgy? Is there any myth associated with the origin of metallurgy? Why did the group choose to work on iron metallurgy? What was the original occupation of the group before adopting smelting and smiting, if any?
6. How do you (e.g. the master smelter) describe the smelting procedure?
7. How do you process the ore, before and during the smelting, such as crushing and roasting? Was roasting essential and was it on the fire or sun?

8. How do you describe furnace construction? What is its height, width at the bottom and top? How many holes are there on the furnace? Is there any special hole designed for extracting, theresidue, the slag, or the bloom? What do you call the slag/bloom? Do you construct or dig a pit under the furnace? If so, for what purpose was it? Do you remember whether the slag was flowing through the base or flowing into the pit prepared for this purpose?
9. Did you decorate the furnace? Why is furnace decorated? What do you call the furnace? Is it a woman or a man? Does the furnace have features indicative of human body parts? What do you call the parts? Why do you call them so? Why do you think furnace is regarded as a female? Did women participate in iron smelting? How often was that?
10. Do you recall how tuyères were prepared? How many tuyères were prepared? How many kinds of tuyères were needed for one smelting or smithing? What do you call the tuyères? Do you call them all by the same name? How long is each tuyère? What do you call the tuyère that penetrates into the furnace? What do you call the other that connects the bellows with the previous tuyère?
11. Do you recall the bellows for smelting and Smithing? Were they different? What is the material bellows made of? Was it made from clay? Was it made from wood? Was it made from skin? How many bellows were used for smelting or Smithing? Who was preparing the bellows-the smelters or smiths or other specialists? For instance, if the bellows were made from skin, was it the tanner's job and what benefits did the tanners get? What do you call it? Why do you call it so?
12. Was there any ritual activity accompanying smelting or smithing? Was it unavoidable requirement? What would happen if ritual aspect were avoided? Do you remember medicinal use of some plants or animals for this practice? Was there any form of sacrifice? If any, for whom was the sacrifice? Where was the sacrifice offered? When was or at what stage was the sacrifice needed? What type of animal was slaughtered? (Goat? Sheep? Chicken?) Is there any **color/sex/age** selection of the animals for sacrifice? Do you think sprits have power to intervene for or against smelt? Which type of sprits needed propitiation? Do you ascribe the failures (unsuccessful smelt, breaking of tuyères, burning of the smelting house etc.,) to intervention by some malevolent forces, envious neighbor, witches/shamans, evils, as well as the angry ancestral sprits?
13. Is there any form of permission or prohibition during and around smelting? What are prohibited? What are permitted? Who are prohibited? Who are permitted? Does age matter? What criterion is expected to be attained by participants? Does this include guests outside the group? Is celibacy from sexual activities a must for every one? Were women allowed to participate in smelting activity? If women were not allowed, how was meal served for the workers? Are sexually inactive women (menopause) or young girls allowed to come to close range of smelting site?
14. Where was smelting conducted? Was smelting conducted in the village? Was it carried out outside the village? If so, what was the reason? Was it to avoid pollution? Was it fear of fire accident? Was there anything like 'producing peacefully' or *waca irraa fagaachhuu* (literally to be distant from noises)? Was there any *modus operandi* for selecting one site over another? Does selection of site have to do with propitiation of ancestral sprits? Was location of smelting site connected with the source of raw

materials (clay, charcoal, construction wood, and ore)? Is there specific space in the smelting house for charcoal, ore, bellows, clay etc? Was smelting conducted inside a house? Was smelting conducted in open spaces? Were these all requirements applicable for smithing?

15. When in the year was smelting conducted? Was smelting conducted during rainy season or dry season? Was it conducted during daytime or nighttime and why? If smelting was to be conducted during rainy season, how was it possible to prepare charcoal? How was it possible to make tuyères and dry them?
16. What do you think was/is your social position *vis-à-vis* other non-handicraft people (the same question is presented for non-artisans)? Who was marginalized? Who was marginalizing? Do you have equal commensality with them? Did you have the same commensality in the past? Did you have similar welfare association together with them? Did you bury your dead in the same graveyard before the churches? Did you marry outside of your occupational/social identity? Why did not you marry outside your identity? What explanation do members of your group give for not marrying from non-tumtuu (smith) groups? Was/is there any ranking (*walcaalmaa*) in the society? Who is regarded at the top or bottom and why? What were/are the manifestations? What were/are the justifications? What was/is the nature of relationship among the craft people? Does/did the smith marry the tanner? Does/did the smelter marry the smith? Does/did the smith marry the potter?
17. What roles other than metal production do/did the *tumtuu* play? Do/did they have ritual place in the society? Do/did they share space with the shamans/*Qaalluu*? Did/do you play key roles in the rite of passage? Who did/do circumcise children of the society? Do/did they/the *tumtuu* provide medicine to the community? What type of medicine? Why does/did the society value their medicine, if any?
18. What does/did the general socio-economic interaction between the *tumtuu* and the non-*tumtuu* look like? How was/is the transaction of products of both sides performed? How was/is the *tumtuu* rewarded for making and repairing agricultural tools? Were/are they paid in kind, cash or rewarded in labor? Were/are they allowed owning land? Were the artisans habiting close to or inside the habitation of members of the mainstream society? Were they living on the edge of the mainstream village, if so why? Were they allowed to keep animals, cattle for instance? Were they taxed for their works? Did they have right to independently negotiate with landowners or the court? Were they communicating through their patrons? Were they tied to patronage?

The purpose of setting guiding questions listed above is to facilitate data acquisition procedures. It does not mean, however, that these questions are standards to be strictly followed. There are a number of challenges in the presentation of the questions to the object of study. The first challenge is the means of presenting them to the audience. For the sake of alternative means, the questions could have been presented either in structured or unstructured method. Structured approach of interviewing is not possible in my study areas in Wollega. It is mainly because the old age iron producers as well as other farmers are non-literate. It is not possible to distribute the questionnaires. In addition, the culture of the society is such that people convey information not in a predetermined straightforward kind of question and answer but in the procedures of work activities and other peer activities such as during drinks and other social engagements. As a result, the questions set above are presented unintentionally such as in work places.

The other challenge is to whom these questions should be presented. This varies from age to sex to from one occupational identity to the other. In this context, it demands that both artisans and non-artisans should be accessed in their own ways. Presenting a certain question to artisans in the presence of members of the mainstream is touchy. Similarly, a non-artisan member would not respond freely about his group's perspective towards members of the artisans in their presence. In addition, the culture of the society dictates to whom the questions are presented. For instance, it is difficult to ask the wives of the smelters or smiths independent of their husbands. While wives are asked in front of their husbands, however, they prefer to limit their words or expect confirmation from the men or say 'what do I know?' Despite the challenges, however, ethnoarchaeology of iron production in Wollega was documented revealing rich information at every stage of the *chaîne opératoire*. Available data was recorded audio-visually.

CHAPTER TWO

2.0. WOLLEGA: NATURAL AND CULTURAL SETTING

2.1 NATURAL SETTING: BACKGROUND

Mesfin Wolde-Mariam (1972) in *An Introductory Geography of Ethiopia* offers the following assessment regarding the geomorphology of the Ethiopian region. According to him, first, during the Precambrian era gigantic mountains were formed through folding and faulting. Then, due to denudation in the Paleozoic period, the gigantic mountains were subjected to lowering and leveling. Furthermore, persistent denudation during the Paleozoic resulted in the transgression of the sea from the southeast to the northwest during the Cenozoic forming a layer of sandstone and limestone along its edges and another layer of sandstones during regression (see Hamrla 1966; Wolde-Mariam 1972: 27). Thus, at the end of the Cenozoic era, three layers of marine sediments were formed over the Precambrian plain in Ethiopia.

The major geological events that were witnessed during the Cenozoic include crustal deformation allowing successive lava flows onto the surface, which gradually led to vertical displacements creating the Ethiopian rift valley, separating the Ethiopia highlands into two main parts (see Barham and Miller 2008; Hamrla 1966). 'In Tertiary age the up warping and sinking movements started and shaped the present forms of mountains and rifts' (Hamrla 1966:441). Wolde-Mariam, in connection, holds that 'whatever metallic mineral deposits in Ethiopia will largely be in association with the Precambrian rocks' (Wolde-Mariam 1972:31). Wollega is one of few areas in the country where Precambrian rocks are exposed allowing mineral exploitation such as platinum, manganese, and iron (Hamrla 1964/ 1966) possible. On the other hand, such vertical and lateral displacement processes have divided Ethiopia into four different Physiographic features (Phillipson 1993; Zewde 1991).

The four categories of the Ethiopian physiographic features include 1) the northwestern highlands and their associated lowlands, 2) the southwestern highlands and their associated lowlands, 3) the central highlands, and 4) the Afar rift system and its block mountains (Wolde-Mariam 1972:37). This chapter is concerned with the description of the study area in Wollega (Figure 2). The focus is on iron production sites in the districts of Abee-Dongoro and Aira-Gulliso in northeastern and southwestern parts of Wollega respectively. Although the following points encompass many regionsoutside the study area, it is possible to roughly situate Abee-Dongoro and Aira-Gulliso within 5° -10° N and 35° -40° E (see Figure 2). It is also within these reference points that other metallurgy sites referred to by different scholars (e.g. Cline 1937; Haberland 1959; Haaland 2004a, Haaland *et al* 2004b; Huntingford 1955; Lewis 1965; Todd 1985) are located. As previously mentioned in chapter one, to be further elucidated in chapters four and five, the areas associated with iron production where ethnographic information is still available are in the watershed of the Blue Nile. Unfortunately, lack of record for metallurgy in northern Ethiopia (including areas immediately north of the Blue Nile) has diminished the possibility of undertaking comparative analysis at wider regional level.

Historically, Wollega encompasses a huge land mass in western Ethiopia immediately south of the Blue Nile and westwards to the Sudan involving diverse climatic zones and

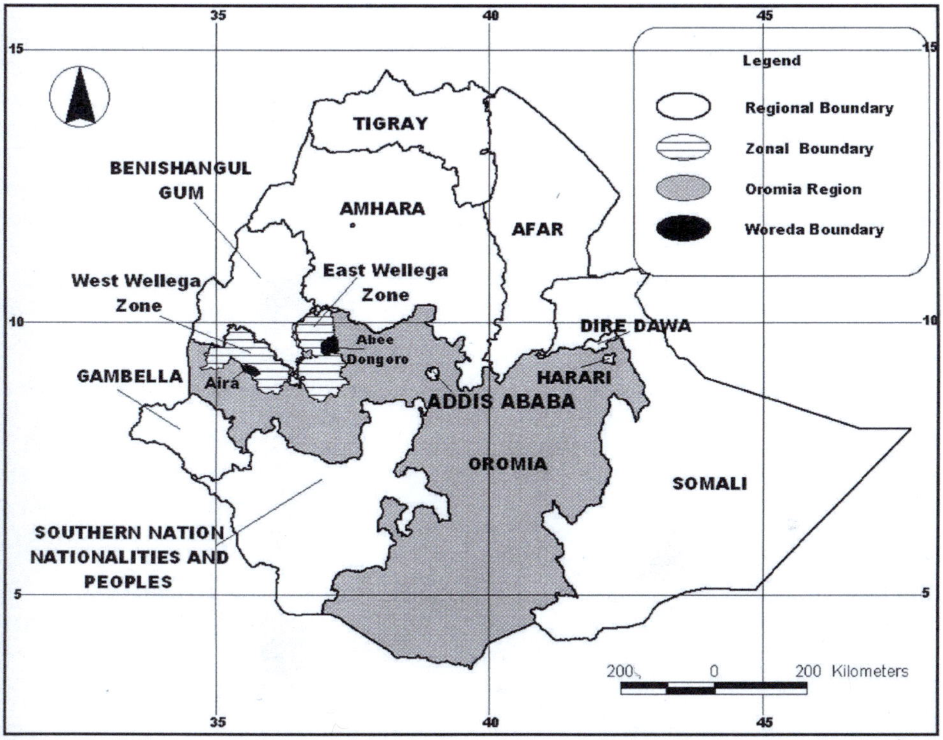

Figure 2 Location map of Wollega according to current administrative divisions. Note areas indicated in the legend as zonal boundaries refer only to the four administrative divisions of Wollega

people with various ways of life. It is situated in the northwestern highlands and associated lowlands that run from Eritrea in the north to Gamo Gofa in the south (Wolde-Mariam 1972, see also Figure 2) forming part of the southwestern plateau (see Phillipson 1993; Zewde 1991). Abee-Dongoro and Aira-Gulliso, the two study areas are situated in Ethiopian northwestern highlands and associated lowlands. In terms of the current Ethiopian administrative division, the study areas are located in east (today sub-divided into east Wollega and Horro-Guduru Wollega zones) and west (today sub-divided into west Wollega and Kellem-Wollega zones) Wollega zones of the Regional State of Oromia, which extends westwards to the Sudanese border.

2.2. CULTURAL SETTING: ETHNOGRAPHY OF WOLLEGA

Various ethnic groups inhabit Wollega (see CSA 1996). During the Imperial period, when Wollega province encompassed western Ethiopia south of the Blue Nile, west of the Gibe and north of the Baro Rivers, in the west extending to the Sudan, the major inhabitants of the region included speakers of the Oromo and Nilo-Saharan languages. Whereas the Nilo-Saharan language speakers of the Berta, Komo and Gumuz communities (see Ahmad 1999, Gidada 2001; Triulzi 1981) settled in the western lowlands bordering the Sudan, the Oromo mainly inhabited the highland parts of the region. However, the ethnography of Wollega should be understood according to changes and continuities that have taken place under successive regimes in the country as well as environmental adversaries outside of it that attracted many immigrants into the region. In this context, one can posit that there was no definite ethnic boundary for such categorization as into Oromo land and of the Gumuz particularly before the present regime. Even under the current regime where language is used as criteria for political boundaries, it fails short of forming a clear physical boundary between various cultures. Many ethnic groups live interspersed in various parts of Wollega.

Such geographic definition of Wollega encompassing regions inhabited by different ethno-linguistic groups mainly applies to the period under the Imperial Regime of Haile Sellassie (pre-1974). This trend has been to some extent maintained under the Derg regime (1974-1991). During the Imperial regime and part of the Derg period, Wollega was subdivided into 49 districts (Waqweya 1988). When the Derg restructured its administrative system in the late 1980s, the region inhabited by the Nilo-Saharan language speakers (commonly known as the Bela Shangul, or Beni Shangul or today Beni Shangul and Gumuz (see Ahmad 1999; Triulzi 1973, 1981 for the various local names) was given a semi-autonomous administrative status independent of Wollega. The part of Wollega stretching westwards to the border of the Sudan, where the main inhabitants are Oromo speakers, has however, remained under the administrative sphere of Wollega. Administrative restructuring and categorization of areas into different zones in Wollega has continued to date.

Since 1991, under the federal government structure of Ethiopia, the size of Wollega administration has witnessed continuous change. This is part of an adjustment to the ethnic line of administrative structure of the Federal Government of Ethiopia. Accordingly, the vast wooded savanna land of the Dhidheessaa valley[6], that protrudes from the northwest direction to divide Wollega into west and east, is included into the Regional State of Beni Shangul and Gumuz (after 1991) on the basis of the settlement pattern of the Gumuz society. The Dhidheesaa valley is part of the region where the previous government has settled (in 1984/85) different communities from drought prone regions of northern Ethiopia. This is in addition to the Gumuz, Oromo and other inhabitants, which were already in occupation

6 The Dhidheessa valley is a vast wooded low-lying landmass extending parts of Wollega north-west wards to the Nile Valley before it empties to the Sudan. It is assumed, according to Wainwright (1954 in Todd and Charles 1978a) that the southeastern most extension of the Meroitic/Kushitic kingdom of the Sudan could have reached and included this valley. Even though, archaeological research is needed to prove or otherwise, if the assumption is accepted at face value, it becomes highly significant for explaining technological variability of iron smelting in Wollega. In his recent book, *The History of Ethiopia*, Saheed A. Adejumobi (2007), mentions Wollega as a transition zone for the expanding food producing communities from western Sudan in about the 4th millennium BC. As I have previously discussed (cf. chapter I), Donald Levine (1974) has given a weight of the antiquity of layers of settlements in this western parts of Ethiopia. With this, if we adopt the theory of itinerant smiths as a model for the spread of iron metallurgy, the technologies in Wollega could be extended for over two millennia since the heyday of the Kingdom of Kush.

of different pockets of the valley. Recently, sesame farming has attracted various groups from different parts of the country especially from across the Blue Nile in Gojjam to settle in the sparsely populated and previously malaria and tsetse fly infested parts of the Angar-Dhidheessa valley. In the past, this river valley has also seen the establishment of state owned mechanized farms. However, the current regime has liquidated and largely privatized the state farm enterprises.

Consequently, the zone in the Dhidheessaa river system towards the Blue Nile has become a scene of confrontation among various stakeholders particularly the 'highlanders' and the 'lowlanders'. The 'highlanders' refers mainly to the sedentary Oromo on both sides of the valley and settlers from northern Ethiopia's drought prone and environmentally degraded regions. Color has also been used to distinguish between highlanders ('red') and lowlanders ('black'). The 'lowlanders' refers to the Benishangul-Gumuz inhabitants who are entitled to administer the land according to the existing political structure. The 'sedentary' highlanders, with their intensive farming culture, have long been encroaching on partially hunter-gatherers and sparsely populated and fertile lowlands of the Gumuz. The Gumuz carry out a slush-and-burn cultivation and supplement their livelihood with hunting and gathering. According to local informants, one of the sources of tension was that the Oromo claimed the area was their ancestral land and regarded the Nilo-Saharan (Gumuz) speakers as new comers. The other is both historical and cultural. Culturally, the 'highlanders' intensive farming practice that involves bush clearing is incompatible with the Gumuz tradition of maintaining the bushes and trees for the purpose of hunting and collecting forest products.

According to Ahmad (1999), the Gumuz began to infiltrate into the areas south of the Blue Nile from the 1840s onwards escaping from Amahra and Agaw slave raiders from the Gojjam highlands. The Gumuz people, whom the locals call derogatorily as *shanqala* (see Hultin 1984), have been victims of slavery and slave raids by the 'highlanders'. According to local tradition, the raids and enslavement of the people continued until it was officially banned in the 1920s and effectively executed during Italian occupation in the late 1930s and early 1940s (on some aspects of the history of the region and period see Ahmad 1999; Hassen 1990; Ta'a 1980; Triulzi 1973; Waqweyya 1988; Zewde 1991). Although the occupation of the Gumuz in the heartland of Wollega is attributed to the mid-19th century slave raids (Ahmad 1999), the settlement of proto-Nilo-Saharan speakers in the west and south west Ethiopian lowland regions is said to go as far back as the 3rd millennium BC (see Fleming 1976; Levine 1974; Todd and Charles 1978a; see also Ehret 1997).

On the other hand, traditions among the Oromo speakers in Wollega attribute their original ancestral land to regions around the Mormor River or the Blue Nile (see Burka 2006; Waqweyya1988; Ta'a 1980), from which many clan names as Abbayyi, Abayyani and the like have evolved. Local histories in northeast Wollega contrast the story that the Oromo occupied the region during the 'population migration' or 'invasion' of the people in the 16th century (for the historical discourse of Oromo history related to the 16th century and the period afterwards see Gidada 2001; Hassen 1990; Jalata 1996; Megerssa 1994; Ta'a 1980; Ullendorff 1960; Zewde 1991). Contrary to written texts of the Ethiopian chroniclers, the Oromo of Wollega claim to have a remote past life along the tributaries of the Blue Nile. As I have noted above, River Mormor (Abbayya) is the area around which the oral history of the

Oromo people spines. One story popular in Wollega about three ancient brothers, presents not only the claim on the ancestral link to areas around the Blue Nile (Mormor) but also the mechanisms for socio-economic differentiation among the people.

2.2.1. TRADITIONS ON THE ORIGIN OF OROMO SETTLEMENT PATTERN: THE STORY OF THREE BROTHERS

According to one of the traditions on the emergence of distinct groups in northeast Wollega along the river Abbaya (Mormor), in the remote past the Oromo used to live in peace in their 'country of origin in Mormor'. They never had quarrels. They agreed on all terms. They campaigned together and defeated their enemies. They hunted and killed big game, which they shared among themselves. They watered their cattle with the *horaa*, (salty waters) in the lowlands of the River Mormor. The difference began with members of one of the families. The tradition continues, as collected by Chernet Waqweyya[7], that an Oromo father had three sons: the first is Boorana, the second Dhandhana and the third Yaabbata. The three brothers used to carry out various assignments under the rule of their father until one day they decided to live on their own. These three brothers asked their father to bless and allow them going far away from his land and starting their own respective life. They decided to go to a region beyond their previous knowledge.

The father of the three brothers blessed them and they departed altogether. On their way to the new adventurous land, they came upon a big river. They saw that the land beyond the river was wet and fertile, suitable for their cattle and for them (as well as the wetland, *dhandhansa*, along the riverbank). The immediate challenge they faced was to find a way to cross the river. They were worried about its depth and the possibility of predators in it. In addition, only the eldest brother knew how to swim. Discussion began on how to cross the river.

Each of the three brothers presented his own plan. Boorana, the eldest proposed that they could cross the river by swimming and walking through it, by *muddying the river*. The two brothers, however, objected to this by stating that not only were they *doonii* (one who does not know how to swim) but they could not trust the river as it could be infested with hunter creatures like pythons and crocodiles. Therefore, they did not want to risk their lives and that of their livestock. Dhanadana instead proposed that since the wetland (*dhandhansa*) is suitable for pasture, they should settle temporarily along the banks of the river until they could devise a secure mechanism for crossing it.

On the other hand, Yaabbata suggested that they climb the trees along the river, and use the branches as a bridge to cross it. Both disagreed with Yaabbata in particular because the option did not accommodate their cattle upon which their livelihood depended. The

7 Chernet Waqweyya is born in Abee-Dongoro district where one of my study sites is located. He is a graduate of history from Addis Ababa University. For many years, he worked as cultural expert in Zonal office in Wollega and documented a number of stories yet to be published. He was one of my field assistants during my research and compiled field data at the end of each visit for me. In addition, his extraordinary rich experience on the oral history of Wollega enabled me to tap from his rich memories. He was also my important informant.

three brothers did not find any common ground from the options. Therefore, they decided to separate by dividing their wealth into three parts. Boorana stayed behind; Dhandhana followed his share of flock along the wetland of the river, and Yabbata went on in search of branches of trees to help him cross to the other side of the river.

Accordingly, Boorana later on entered the river and crossed it with his share of livestock. This changed the color of the river into muddy water, *booruu*. He settled on the other side of the river, began to till the land, possessed it, and prospered. In the end, his descendants came to be identified with this name and were those who became settled farmers. How is this story viewed in line with the place of the name in the context of the Oromo of Wollega? It is intriguing apparently that the story radically diverges from its present connotation. At present, the term Boorana refers to the dawn, the time when the dark night is leaving the place for morning light, from which the adjective 'people of the morning' (see Hultin 1984; Knutsen 1967; Megerssa 1994; Triulzi 1996) is believed to have emerged. The 'people of the morning' (see Megerssa 1994) is derived from the verb *Boru* meaning the morrow. In the context of the story, however, the name could be derived from another root *booruu*, literally means muddy water/the color of sky at dawn. What does muddy water look like? What does the dawn look like? As far as the color is concerned both dawn and muddy water is similar, i.e. dawn is the transitional zone between dark and light- of reddish color as well as the color of the water, which is mixed with mud. In Wollega, the eyes of a Booranaa person are equaled with *booruu*, muddy water, or the color of dawn (see Bartels 1970).

The saying '*kan booressee ce'he'*, literally 'those who muddied the river and crossed it', is associated with those who claim to belong to the Boorana. In other words, the Boorana group is associated with the color of the soil. This probably implies that the Boorana were the occupational/social class of the Oromo society who worked on the land or who were sedentary farmers and that, due to their eldership (see Burka 2006; Megerssa 1994) they were entitled to its ownership. Today, however, the name Boorana stands for a chiefly and ritually higher stratum of the Oromo society in Wollega.

On the other hand, Dhandana and his descendants were said to have continued to live along the wet banks of the river, called *dhandhansa*. They mainly practiced animal husbandry. *Dhandhansa* is a kind of lawn where animals are unleashed for grazing and watering. The term Dhandhana, which the tradition attributes to have been derived from *dhandhansa,* is said to represent the identity of his offspring. However, there is no ethnographic evidence such as clan or class names, which are associated with cattle herders in Wollega. This is probably due to the present absence of pastoral life in Wollega. On the other hand, there is no group solely depending on animal husbandry in Wollega. The people of Wollega are mixed-farmers. For instance, the highland parts of Horro in northeast Wollega is known for diversified species of barley and wheat farming and horse and sheep rearing (Waqweyya 1988). The presence of tens of species of wheat and barley might be evidence of long period of experimentation and adaptation of the crops to the local environment.

Meanwhile, the third brother by the name of *Yaabbata* departed from the group and started a separate life. *Yaabbata* literally means 'he who mounts'. Nevertheless, as I shall explain below, it connotes a person of mixed family origin. The story continues to hold that during his attempts to cross the river by climbing and using the tree branches, Yaabbata became accustomed to collecting forest honey. He liked it and made his life in the forests. He also hunted animals and gathered plants to supplement him with additional food. He also built on hunting tools. His descendants became asmany as the rest of his brothers. They became craft workers such as *gaagurtuu* (beehive makers, hangers), *tumtuu* (smiths), *duugduu* (tanners), and the like.

Contrary to the story of the 'climbers', the term *Yaabbata*, is associated with radically different meaning in the context of the present ethnography of Wollega. The name *Yaabbata* is still applied to various individuals in the region and is a derogatory term. However, no clan in Wollega is associated with this identity. It has no lineage basis. The present meaning has nothing to do with craft workers (smiths, tanners or potters or even beehive makers and honey collectors) as its workers belong to different social strata. Instead, it has a connotation of those who have a mixed family background-between the so-called the *Boorana* and their slaves[8] (i.e. who were born in illegal mating between an Oromo and his female slave of mainly different ethnic origin).

However, some scholars suggest that the *Yaabbata* belonged to groups of people who were adopted by the Oromo into their groups (see Hultin 1984) and were the ones who served as its cavalry forces. This assumption is consistent with the two Oromo institutions of adoption namely *guddifacha* and *moggaasaa* (see Hassen 1990). These two institutions were claimed to have been instrumental for the Oromo 'invaders' to easily occupying vast Ethiopian regions in the '16th century' (Hassen 1990; Megerssa 1994). The Oromo adoption of others at individual and clan level allowed the new members to claim Oromo identity through their respective parental clans (Burka 2006).

According to Hassen (1990), such a mechanism is in contrast to the culture of the Ethiopian highland Amhara whose conquered people become slaves and explains the fact that the non-Oromo inhabitants of various regions have been easily assimilated to them. Using a similar model, Hultin (1984) argues that the Oromo (in Wollega) used as cavalry forces some of the adopted from various societies, called *Yaabbata*, meaning those who 'mount horse'. Today, however, horse riders in Wollega are not known by this name. On the other hand, horse keeping and riding is regarded high in Wollega as opposed to the meaning that *Yaabbata* has in the society. In addition, horse rearing is not a distinct specialization by certain groups but a common practice of farmers in the different highland parts of the region.

The presence of such a tradition and its context enables us to enrich our perspective of how the Oromo people see their past. It is evident to show how people conceptualize and spine webs of significance to their past through tales. It does by no means imply that the Oromo settlement pattern, along occupational identities in Wollega, has been structured purely according to the story of the three brothers. It is only an alternative to the meta-narrative that the Oromo came to the region in the 16th century in 'waves as pastoral hordes' (see

---

8 Bartels (1970) has a detailed discussion on the status of former slaves among the Oromo society in Wollega. The fact that the Oromo kept slaves is another evidence of the presence of stratification in the society as well as any of the deep-rooted culture of highland societies in Ethiopia. That the Oromo allowed assimilation only shows that their culture was different from their neighbors such as the Amhara. This, however, does not mean that the Oromo were communal and as result allowed equal commensality to the assimilated individuals or groups (see for such discussions Triulzi, 1996)

for critics Jaleta 1996; Megerssa 1994). However, whether the story of the three brothers represents the period before the 16th century or after is still open for discussion. One thing that comes out of traditional story is that the Oromo have their own perspective angle of how they were differentiated into different strata since time immemorial. In general, our knowledge of environmental and cultural setting of the two study areas would contribute much to our understanding of the place of artisans in Oromo society.

Abee-Dongoroo-population size according to major ethnic groups (CSA 1996:187)										
Item	Urban and rural			Urban population			Rural population size			Ref
	Total			Total			Total			.
	Male	Female		Male	Female		Male	Female		
All persons	39042	19382	19660	1000	496	504	38042	18886	19106	
Agaw	53	30	23	-	-	-	53	30	23	
Amhara	19791	10059	9732	111	61	50	19680	9998	9682	
Kambata	7	5	2	-	-	-	7	5	2	
Werji	1	1	-	-	-	-	1	1	-	
Oromo	18796	9083	9713	888	434	454	17908	8649	9259	
Tigray	384	197	187	1	1	-	383	196	187	
Others	10	7	3	-	-	-	10	7	3	
Aira-Gulliso district population size (CSA, 1996:184)										
All persons	80321	39542	40779	9043	4439	4604	71278	35103	36175	
Amhara	1399	754	645	417	223	194	982	531	451	
Arborie	14	5	9	6	1	5	8	4	4	
S.bet Guragie	57	37	20	57	37	20	-	-	-	
Silitie	62	42	20	62	42	20	-	-	-	
Oromo	78677	38645	40032	8442	4108	4334	70235	34537	35698	
Werji	2	2	-	-	-	-	2	2	-	
Tigray	49	26	23	16	8	8	33	18	15	
Others	61	31	30	43	20	23	18	11	7	

Figure 3 population sizes according to ethnic types in Abee-Dongoroo and Aira-Gulliso districts. Data is obtained from the 1994 population census result published in 1996 by the CSA, FDRE

2.3. ABEE-DONGORO: NATURAL AND CULTURAL CONTEXT

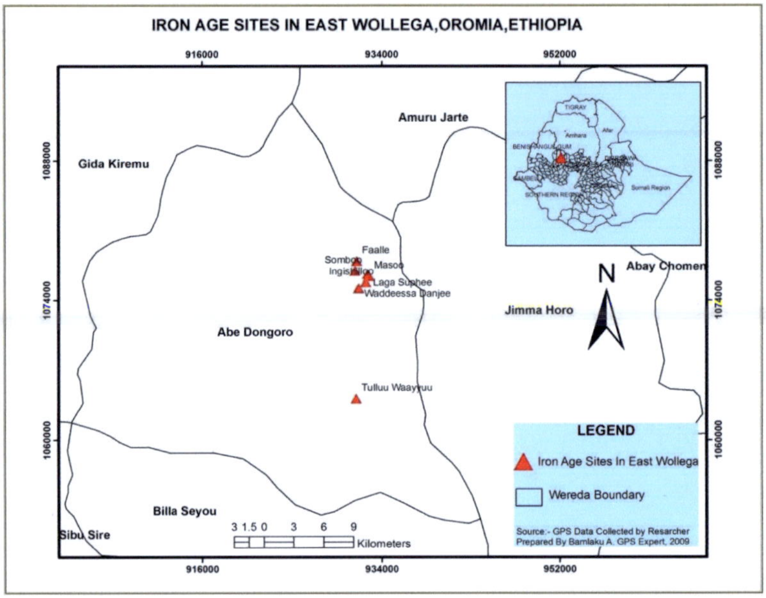

Figure 4 Relative location of iron mining, smelting and smithing sites in Abee-Dongoro district northeast Wollega: the points were taken from different villages consistent with the settlement pattern

The official socio-economic profile of East Wollega Zone, describes Abee-Dongoro as physically characterized by rugged topography. The main peaks in the district are Borge (2592 m), Wando (2532 m), and Bijjit (1972m). Tulluu Waayyu, the district administrative capital is situated on the plateau at an altitude of 2361m according to GPS point taken at St. Gabriel church in the town (Burka 2006). Iron mining and production centre is located 20-25 km away from the capital descending to about 1700 m, a GPS reading taken at clay extraction site called Laga Suphee (Figure 4). Most parts of the district are located in broken lands overlooking the low-lying flanks of the Hangar-Dhidheessaa valley.

Abee-Dongoro is situated in the north central part of northeast Wollega. In the north, Amuru Jartee district bounds it, in the east Jimma Horro, in the west and northwest Gidda Kiramu and in the south Biila Sayyo district bounds Abee-Dongoro (Figure 4). The districts derive their names from minor or major clans or a combination of both. It is sometimes easy to demarcate between lineages, as closely related ones uniformly inhabit in defined territories. Valleys, creeks and other natural features can separate lineages. Theoretically, a common land for pasture or forest for hunting can be left between the lineages. This probably explains why Wollega is still forested in contrast to neighboring Gojjam across

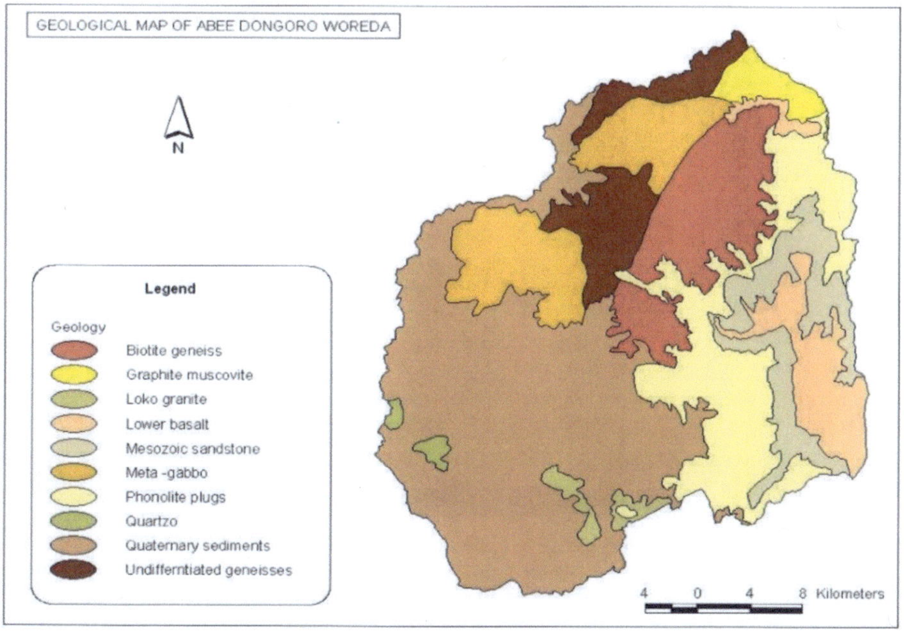

Figure 5 Geological map of the district of Abee-Dongoro

the Blue Nile. Travelers in the 1840s, like Antoine D'Abbadie, have clearly noted the Oromo for their protection of the natural environment as opposed to their neighbors north of river Mormor or Abbayya (Kanno 2007). Today, population size has increased and many marginal lands in the past are now put under cultivation. The high demand for oil seeds such as sesame cash crops has led various people including from north of the Blue Nile to migrate and open up the previously inhospitable lowland (*gammoojjii*) areas in the district (see Figure 3 for population size in Abee-Dongoroo district *vis-à-vis* Aira-Gulliso). Environmental degradation through deforestation of the sesame belt of the *gammoojjii* parts of Abee-Dongoro district by the settlers is profound. Combined with the recent population pressure in the highland Dongoro and the realizationthat sesame farming is much rewarding, land grabbing and confrontation with settlers from the Amhara Region of Gojjam has recently been intensified.

On the other hand, the regional government's profile lumps the district as being generally dominated by temperate climate. However, it can be categorized into three locally understood classifications: *Baddaa* (cooler uplands), *Badda-Daree* (mild midlands) and *Gammojji* (hot lowlands) constituting the cold upland *Baddaa* >2000m, the *Badda-Daree* moderate midland between 1400 and 2000m and the *Gammoojii* areas below 1400m (see Megerssa 1994:29-30). The local traditional categorization is generally established according to relative temperature of the localities. Local knowledge of what kind of crops and animals are suitable to each zone is identified and plants and animals[9] are cultivated and reared accordingly.

According to data from the district administration office (a chart on the wall[10]), the total area of Abee-Dongoroo is 109,209 hectares. Out of this, *Baddaa* constitutes 2,981 ha, *Badda Daree* 11,941ha and *Gammoojjii* 94, 280 hectares (Burka 2006). Depending on this data and other written documents about the region, over 86% of the total area of the district of Abee-Dongoro constitutes an area that has never been suitable for human habitation and animal husbandry in the past (see also Waqweyya 1988).

According to Waqweyya (1988) and my oral informants in the district, the problem was due to malaria and animal disease caused by the tsetse fly. Cattle are only seasonally taken into the lowlands for watering them with salt waters (*Hora*), the activity called *Hora obaasuu*. Elders remember that *Hora obaasuu*, which was conducted by the young males, used to take place immediately after the summer time when farming and sowing were over. This activity is no longer functional partly due to animal diseases and partly due to cultural changes over the years. According to elders, the regular campaign to the salt water, *Hora* areas, does not seem to have taken place in the last 40-50 years.

Abee-Dongoro is presently under transformation, on other grounds. The location of the district close to the Blue Nile valley across which land problem is acute in Gojjam, the availability of much unoccupied land in the district, the opening of highway route connecting Wollega and Gojjam facilitating population interaction, and many other reasons can be cited as factors of dynamism accelerating change in the district. Particularly, the new highway opened in the late 1980s along the traditional caravan route between Wollega and Gojjam across the Blue Nile has intensified population interactions.

---

9 A local story narrates that sometime in the past, goat, and sheep made an intimate friendship. As their friendship strengthened, they agreed to visit the habitation of each. Thus, goat was invited to the house of the sheep. It was a cloudy and rainy area. Sheep asked his friend to go out grazing. Nevertheless, the goat was not up to walking out into the rainy cold outside. He proposed to stay until the sky cleared. It did not. Goat called loud and asked the sheep 'is it raining'? Sheep answered 'it sprinkles, I still graze'. The whole day, the same question and answer. The story concludes that goats and sheep must be kept in different ecological zones and in the pens where goats are kept, flat stones, logs of wood are kept for them to stand on to avoid wet ground.

10 Abee-Dongoro is relatively a remote district when compared to other similar districts mainly because of its rugged physical features. As a result, road transport from sub-provincial capital at Shamboo, less than 50 km distance is only possible during dry seasons. Multitude of river streams but particularly the Garchii River seems to have barred connection with the sub-provincial administrative capital. Occasionally, from former state farm lowland of Hanagar Dhidhessa valley, some trucks have ascended to Abee-Dongoro. This was only two times a week-on Tuesday and Saturday. The year 2004/2005 seems to be a year when some local data was made available particularly owing to preliminary population census and land estimation of the district for the purpose of the May 2005 national election. The various district offices of Abee-Dongoro were not equipped with archival materials conveying information on the history of the people in the region. The chart/graph on the wall inside the office of the district administrator was the only written information discussing the socio-economic statistics of the district.

According to locals, in less than fifteen years during the current regime, tens of hundreds of inhabitants from Gojjam crossed and settled deep in the wooded lowlands and engaged in slash-and-burning agriculture of sorghum, maize, and recently sesame. Claims to the lands by the native surrounding population has led to many bloody clashes and the regional governments of Oromia and Amhara have intervened in the resettlement of the immigrants. In response to the problem of animal disease, sesame farmers in the district have recently introduced farming by donkeys. Flocks of donkeys are increasingly crossing the Blue Nile from the north whence traditions hold that beasts of burden were imported since immemorial time originating in the neighboring Sudan.

The district is generally wet, from which various rivers forming the tributaries of the Nile originate including Arjo, Lugo, Indode, Garchi, Demonu and Gimbi (see Burka 2006; Waqweyya 1988). According to the district's profile, the soil types of the district are generally Orthich Acrisols and Nitosols. Dystric Nitosols are the dominant soil types and important for cultivation due to their high water retention capacity. The variety of formations can be the result of the nature of the local geology of Abee-Dongoro (see Figure 5).

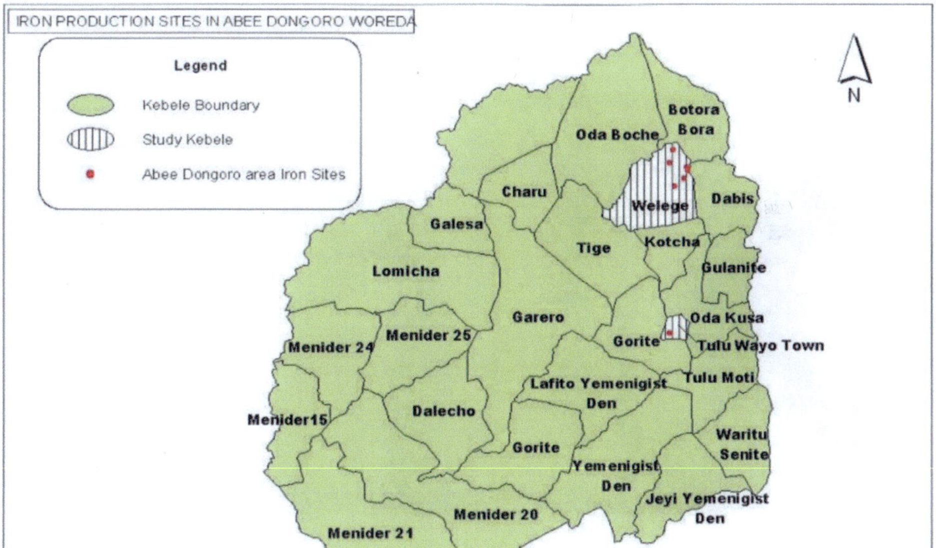

Figure 6 Relative location of iron producing sites in lowland Dongoro at about 20km distance from administrative capital

Stratified metamorphic rocks (Figure 5) characterize the local geology of iron deposit and production area. Particular local characteristics are Graphite[11], Muscovite[12], and Quartz schist[13] ranging from light grey to slivery grey and of fine grained strongly schistose. In addition, there is lower basalt: olivine[14] -pyroxene[15] -phyric basalt with some aphanitic[16] layers (Figure 4).

Faallee, the iron ore-mining site (see Figure 5) is of Biotite gneiss. Iron ore has been mined and smelted from this district and was exchanged in great quaintly for *amole* (salt bar) at Assandabo (Qawoo) market in Guduru (close to river Abbayya in the northeast of Wollega) as far back as the 19th century (see Kanno 2007; Triulzi 1973; Waqweya 1988 and oral tradition). Its bars or fragments have also been used as a medium of exchange with other goods in the local markets. Old women (widows) who regularly visited smelting sites and collected fragments used to sell them to smiths and lived on them.

People commonly believed that individuals easily got rich if the money obtained from the sale of iron was used to buy a heifer, *goromsaa*. According to elder informants, during the Great Famine (1888-1892), Abee-Dongoro's iron mining and smelting served as life-saving refugium for many people from north of the Blue Nile (Abbayya, Mormor). Nevertheless, there is no information on the genealogy of a group near iron production sites claiming descent from Gojjam. Elders attributed the absence to the fact that immigrants only stayed there until the bad days have passed. Informants added that in the first half of the 20th century, it was still common for migrants from Gojjam to come and work seasonally on the mines, collect money, and return.

Informants tried to justify the maintenance of the tradition of labor migrants of hired *Gojjame* young farmers on agreement of one-third or one-fourth (known as *siso/Irbo* respectively) share of the produce of the plot they farmed. According to them, very few permanently stayed, while the majority returned after they collected enough money during their years of stay in the client-patron relationship. When Antoine D'Abbadie visited and stayed among the Guduru of northeast Wollega in the 1840s, this part of Wollega was defined as beautifully green attributed to the Oromo tradition of protecting forest (see Kanno 2007) and was under various regional republics whose administration was based on the age-grade *gada* system.

The wet environment also favored patchily distributed forests mainly along the Garchi[17] river which were inhabited by lions, buffalos (are no more in the upper part of the valley)

---

11 used as a lubricant and as a moderator in nuclear reactors e.g. black lead,

12 a colorless or pale brown mica with potassium

13 any metamorphic rock that can be split into thin layers

14 A mineral consisting of magnesium iron silicate; a source of magnesium

15 any of a group of crystalline silicate mineral common in igneous and metamorphic rocks

16 fine-grained homogeneous rock (such as basalt) containing minerals undetectable by the naked eye

17 Garchi is a name both for a low lying valley and river system where tradition holds that a seasonal campaign for big game used to be conducted by adventurous groups from different parts of the highland Horro.

and other small and big wild animals (Waqweyya 1988). According to the tradition from the highland parts of Horro-Guduru, Abee-Dongoro used to be one of the localities where hunters from different parts of the Horro highlands made seasonal campaigns for big game. The vegetation of northeast Wollega is generally categorized as upland grassland (Wolde-Mariam 1972) with localized tropical highland forest coverage.

The varied climatic system of the district (due to its rugged characteristics) seems to have influenced patterns of population settlements and enabled diverse economic activities in the Abee-Dongoro district. Abee-Dongoro can be categorized into two settlement patterns based on its physical features and occupation by different super-clans: Abee and Dongoro. The Abee part of the district relatively refers to the mostly cold highland parts whereas Dongoro constitutes parts of the highland, midland, and the rest of hot lowlands. In the Abee area, the main crops produced include wheat, barley, lentils, beans, and peas. The animals reared are mainly highland long tailed Horro breed sheep and horses (Waqweya 1988). In the Dongoro area, other major farm products include teff, sorghum, finger millet, and maize. Zebu cattle (Bos indicus) -known locally as *Botee*[18] (Waqweya 1988) are typical animals of the Horro-Guduru parts of Wollega as well as in Abee-Dongoro.

In the hot lowlands, crops like maize, sorghum, and sesame are cultivated. The dominant animals are donkeys and mules, which are imported from areas north of the Blue Nile (Waqweya 1988). In the map (Figure 6), for instance, villages named with numbers refer to new sites where immigrants recently settled and cultivate lowland crops mentioned above. In addition, other crops produced here include coffee, citron, ginger, and cardamom. In the past, it is held that the hot lowlands provided the main source of forest honey. Forest honey used to be collected and still the tradition is maintained in the lowlands as well as beehives are hanged on deciduous trees such as oaks, which are dominant in the ecological zone. On the other hand, some localities such as Laagee and Tigee neighboring iron production sites are known for their coffee farming (see Figure 6).

However, people who were recently resettled from drought prone areas of eastern Ethiopia have begun to rear goats and grow Chat edulis, a mild stimulant plant, which is commercially rewarding. As a result, the installation of veterinary services recently, the lowland settlers have also begun to rear other domestic animals. Apparently, also the further opening up of the wooded lowlands has reduced the impact of tsetse fly (genus Glossina), the bloodsucking African fly that transmits sleeping sickness.

Walagee is one of the peasant associations or *ganda* in Oromo language, the structure of which often corresponds to lineages of Abee Dongoro. The iron production centers of Walagee are located about 20-25 km to northeast of *Tulluu* Waayyuu, the administrative capital. *Ganda* Walagee is situated in the lowland part of the Garchii river valley and relatively hot and humid part of the district. It is a transitional zone between temperate midlands and hot and humid lowlands. The altitudinal difference between the district capital and the iron production locality is about 600 meters and more 400 meters minus would take one to descend towards *gammoojjii* part of the Hangar-Dhidheesa river valley.

18 *Botee* refers to long horned livestock and are typical of Horro-Guduru and the rest of Wollega cattle.

Figure 7 Note a slash and burn farming encroaching to the forest. Charcoal is regularly made for iron production activities from this part. It is close to the village of smiths (Masoo).

Nevertheless, some areas are located in the lower altitude, but because of forest coverage such as *Mengist Den*, which literally means 'government forest' (Figure 6) and high number of river streams, Abee-Dongoro is generally wet and humid. Over 30, 000 hectares of Chaatoo forest in the Garchii river valley also contributed to the fertility and wetness of the district (Figure 7). Despite low-lying location, because of the influence of the peaks surrounding it, Walagee is within the *Badda- Daree* (midland) climatic zone.

2.3.1. WALAGEE AND ITS ENVIRONS: THE WAY OF LIFE

Abee-Dongoroo is well known for its natural resources. Walagee, the study locality is still fairly covered with wooded forest[19] (see Figure 7) that partly explains why iron production is concentrated in these lowland areas than elsewhere in the district. This (Figure 7) is not

19 The picture is taken from the vicinity of the Masoo hamlet where over 31 heads of households of smiths live. Smiths are also farmers, who cultivate different crops and rear animals. Despite encroachment by farmers, the surrounding area particularly along river streams, which are abundant, is still under forest coverage. The forest has provided and is still a source of charcoal for smelting and smithing. Various hard woods provide choices for the smiths (cf. chapters III&V).

a formally protected forest but it seems that the mystery of its survival must stem from the long based rich tradition of respect for the environment by the local inhabitants. However, change is inevitable since in the photo taken in 2005, encroachment by farmers is evident.

On the other hand, the forest called *Chaatoo* (some over five kilometers from the research center) is a source of humidity and rainfall for much of the year. This is in addition to forest coverage in different pockets of the district (as in Figure 7). The *Chaato*[20] forest is situated between the Abee-Dongoro district on the one hand and Horro Jimma and Amuru Jaarte on the other. One important natural resource of Abee-Dongoro is an iron ore (hematite), called *Gordana Sibiilaa*. *Gordana* stands for the ore while *sibiila* means iron or metal. The Oromo term for iron is also *sibiila gurraacha*, black metal (see Wainwright 1942b for various terms).

This ore is located in Walagee, one of over 25 lower administrative units of the district (Figure 5). Walagee's elevation varies between 1778 m at *Laga* suphee (Clay River), clay extraction site, and 1820m around the hamlets of ironworkers and farmers. Hamlets that are important in connection with iron working in Walagee included Faallee, Sandaaboo, Somboo, Waddeessa Danjee, Ingishilloo, Jawaj and Masoo (Figures 4 and 6). The settlements are in accordance with the closest lineages as well as occupational identities. For instance, Faallee, Waddeessa Danjee, and part of Jawaj inhabitants dominantly belong to Malkee lineage. The iron mining sites and many of the smelting sites are located within reach of these hamlets-mainly at Waddeesa Danjee. Accordingly, one of the Malkee sub-lineages seemed to have controlled the iron mining sites, its mining activities, as well as the smelting occupation. Members of the Yamalagii clan occupy Somboo and Masoo hamlets predominantly. These hamlets are inhabited by the smiths who once worked on the blooms and are still forging scrap iron.

Coffee is an important cash crop in the Walagee area. Inhabitants of the *gandaa*, including the smiths plant coffee on their respective farmsteads as well as in common forest areas along river streams. Maize, sorghum, and finger millet are major staple crops produced in Walagee. *Teff* is also produced but not in great quantity like maize, finger millet, and sorghum. The inhabitants of Walagee are mainly Oromo (the *tumtuu* also count the same origin) but for some non-Oromo priests and immigrants who live around the Orthodox Church. Iron forging communities (the *tumtuu*) largely populate Walagee in separate hamlets of Somboo and Masoo.

Ironworking is an exclusive male occupation. There is no ethnographic evidence on the occupation of the wives of the smiths' wives as potters in Walagee area. No single potter is recorded in these hamlets as opposed to in the past when, for instance, the mother of the master smith, *Obbo* Darbush Kumsaa, was potting. However, in the neighboring *ganda* Dabbisi, many women are still working as potters. Nevertheless, in terms of house structures and other ways of life there is no significant difference between the smiths

20 Chaatoo (Caatoo) refers to a spirit of the forest. The spirit allows hunters to kill big games. If Caatoo did not allow, even any perfect hunter cannot kill beasts. As a result, rituals to appease *Caato* are conducted during the hunting campaigns. The *tumtuu* who make the spear point or revitalize the gunpowder are conducting the secret role of appeasing the *caatoo*.

and the members of the mainstream including smelters. The hamlets (Faallee, Somboo, Ingishillo, and Masoo) are separated from one another by small streams forming, for instance, a separate habitation areas between the smelters (who claim to belong to the mainstream) and the smiths, who are being 'distanced' (see Burka 2006, 2008, 2009) from equal commensality with the rest of the population.

Somboo and Masoo hamlets are inhabited by 31 households (Burka 2006), all of which claim Oromo descent. Three of these households belong to ex-slave families (although now they speak only Oromo language) while twenty-eight of them belong exclusively to the *tumtuu*/smith mainly from Yamalagii lineage. On the other hand, the Malkee farmers and / or smelters dominantly inhabit in other localities such as Waddeessa Danjee, Jawajj, and Faalle. The Malkee lineage is ultimately linked to the Gutaa upper lineage whose tradition holds that their members were warriors whose spearhead came from northwards close to the Blue Nile/Abbaayya. They are widely spread in northeast Wollega as well as in the southwest. There is a tradition among the Yamalagii that holds that their ancient origin was also around the Blue Nile in northeast Wollega. They claim that originally their ancestors were cattle herders. They were always led by the decision of the bull that led the flock. According to them, people with the same lineage name in northwest Wollega close to the Nile (Abbayya) were their elders (ancestors) who were left behind.

The *tumtuu* live on their craftwork as well as farming and rearing animals. As a result, they are well off economically. Craft working, in the present, is conducted on part-time basis, however. The major source of income is obtained from farming various crops and by rearing animals. Unlike in other areas where pockets of smiths depended on their patrons for their livelihood, in Walagee, either due to their high number or their importance in providing tools sought by highland farmers, or because the lowland was not favored for settlement in the past, we do not have a story connected to land problem. The grandfather of one of the smiths, Obbo Darbush, who volunteered to reenact the smelting, was a *Balambaras*, a noble title given to the upper class of the mainstream. This probably happened after Amhara occupation of the region in the later part of the 19th century or even following the region's subordination to the rule of Nigus Teklahaymanot of Gojjam since the late 1870s. Donkey and mules as well as cattle and goats make important pack and farm animals in *ganda* Walagee of Abee-Dongoroo district.

Slash-and-burn has been a major mechanism of farming activity in the past (Waqweyya 1988). According to local tradition, in the past, much of the work was not carried out by ox-drawn plough. Today, ox-drawn plough farming is used mainly in open areas side-by-side with slash and burn system. The major iron tools produced at present and in high demand at the forgers' hut (Burka 2006) are different sizes of bush knives (*gajamoo* and *galadoo*). These two tools are needed for slashing or preparing farm fields and clearing coffee plant areas from weeds.

The present high demand for mainly *gajamoo* and *galadoo* (big and medium size curved bush knives respectively) is mainly due to the increasing encroachments towards the formerly inhospitable hot lowlands of the district for different economic activities. This is contrary to what is known from oral history about the type of iron tools that were in high demand in the area some fifty years ago.

According to informants and other sources, in the past, when Walagee's iron was part of trade items and in high demand in distant areas in north Ethiopia (Ahmad 1980; Triulzi 1975; Waqweya 1988), iron bars and plowshare were produced in large quantities by the Walagee smiths. It is clear from the story that iron production (raw iron and plowshare) was much for trade outside the district to the high lands of Horro and Guduru and to areas north of the Blue Nile in Gojjam but less for local use. In contrast to the past, today the demand for iron tools is from the local. Before setting out for other activities, smiths got busy in the mornings and after farming in the evening to produce as many tools, bush knives in particular, as possible. The artisans or other intermediaries sold various tools particularly the bush knives either at weekly market in Tullu Waayyu on Saturdays or at local markets in the villages at Botoro Booraa, north of Walagee (see Figure 6).

Despite their better economic status, however, there is lack of equal commensality between the smith/*tumtuu* and members of the mainstream community including the smelters/ *buuftuu*. An examination of their marriage history provides that out of 28 household heads in the hamlets of Masoo and Somboo where the *tumtuu* concentrated, only two of them could forge exogamous marriage. According to my informants from Malkee lineage iron smelters (e.g. *Obbo* Moosisaa), one of the *tumtuu* descendant, who during my research was head of the *gandaa*, married to a daughter of former smelters, but was reportedly resented. There is no much difference with regard to the status of the *tumtuu* in southwest Wollega except for similar incidents as I have mentioned above.

Nevertheless, a number of other approaches including genealogies and lineage relationships showed that marginalization of the *tumtuu* occupational and social identity is much of superficial one (Burka 2009). I will later on discuss manifestations of marginalization of the *tumtuu* based on stories from both the artisans and the mainstream groups (cf. particularly *chapter VI*). A similar time interview of the marriage relation of the smelters (*buuftuu*) provided a rather different picture from the position of the *tumtuu* (Burka 2006). The smelters had exogamous marriage and mainly went to the extent of counting marriage relationships with many of the chiefs in different parts of the region outside the immediate district. I will expand on the discussion of the social position of members of the artisan groups under the topic addressing ranks among the Oromo below.

2.4. AIRA-GULLISO: NATURAL AND CULTURAL CONTEXT

The administrative restructuring of Wollega as elsewhere in Ethiopia is still a continuous process. Wollega has previously been a single administrative province, known as *Teklay Gizat* before the Imperial regime was replaced. In the last 18 years of the new government, a number of categorizations have been made to Wollega. Previously under the Derg, the military regime and under the incumbent regime, it has been sub-divided into two administrative zones: East and West Wollega zones. Nevertheless, since some four years ago, it is still undergoing sub-divisions making the number of zones four (i.e. the east one into two and the west into two).

Figure 2 indicates such a current division of Wollega. Part of this restructuring has also affected the former Aira-Gulliso district, which has previously been administered from the town of Gulliso.

Figure 8 an old furnace builder, Obbo Waratii Fayyisaa, and tuyères marker from Yubdo district standing on ancestral smelting site in a corn field. Note in the far background vestiges of natural trees and bushes. From where the furnace builder stands the elevation immediately descends to a nearby stream where still, some canopy trees provide sheds to coffee[21] plants.

As a result, recently the district of Aira-Gulliso was further divided into two parts-Aira district and Gulliso[22] district.

The research on iron production in southwest of Wollega was conducted in accordance with previous administrative structures (e.g. Burka 2006, 2008, 2009). For the purpose of focus, I shall concentrate on its present structural setting for Aira district part of the reason of which was that most of the localities for the study sites are situated within this

21 Coffee farming might have been the major reason for survival of indigenous plants. As coffee is cultivated, near streams and under sheds of trees, along these we find plants that have sustained human destruction. In the rest of areas close to other farms and villages, the dominant trees are foreign eucalyptus species.

22 Gulliso is located about 20 km north of Aira town. In the past, when the district encompasses areas around the two towns, Gulliso served as administrative seat of the district administration. My letter of permission to conduct research in the region, written from the west Wollega zone culture and tourism office in 2005 has addressed to Aira-Gulliso and the capital as Gulliso. Nevertheless, since 2007, I found out that a new administrative structure was being developed sub-dividing the former district into two parts with their respective administrative seats at Aira and Gulliso towns.

new district. However, there are archaeometallurgical sites in Gulliso district including the well-known mining sites called Caaggoo and Warra Qaalluu (see Hamrla 1966) as well as smelting sites. On the other hand, the division of the area into two districts does not mean that the people are culturally distinct from one another in the past. For instance, some former smelting sites and its workers are outside the recent Aira district and are situated in Yubdo district (see Figure 8) although the settlement is only separated by a few hundred meters.

Unlike Abee-Dongoro district, which is characterized by various mountain peaks and broken lands, Aira-Gulliso district consists of hot and uniform low-lying plains. It is part of southwest Wollega, where the land seems to have been over utilized and parental rocks are exposed (Hamrla 1966). As a result, there is a trade of cereals (beans, peas, *teff* etc.) from northeast Wollega, which is a surplus productive part of western Ethiopia. The southwestern Wollega region's main agricultural produce (despite many types are produced) is sorghum. Durra (sorghum), together with iron, has been traded from this part of the country to the direction of the Sudan as far back in time as the 19th century (see Ahmad 1999; Triulzi 1975). The environment of southwestern Wollega around Aira-Gulliso is suitable for durra and finger millet crops. The other backbone of the economy of the district is coffee. Iron is also another source of income particularly around the *tumtuu* concentrated hamlets in Jaarsoo Abayyanii. Few of natural deciduous canopy trees and some hardwoods are available mostly along streams interspersed with coffee plants (Figure 8). Generally, the average altitude of the district is below 2000 m above sea level.

According to available sources, southwestern Wollega is one of the few Ethiopian regions where Precambrian rocks are exposed. 'The pre-Cambrian basement complex must be considered as the potentially most favorable environment to contain primary high-grade ore, but also secondary low-grade ore' (Hamrla1966: 441). This endowed it with various types of minerals. Aira-Gulliso and its surrounding districts are sources of major minerals such as iron, gold and platinum. The platinum mining site at Yubdo, which was leased to mineral explorers such as Alberto Prasso in the early 20th century, is a case in point (see Cerulli 1933). Yubdo area is also source of good quality iron ore at a locality called Dimma (thus *gordana dimma*), from whence traditional smelters used to mine and whence they brought for the smelting reconstruction project of 2007. However, Aira-Gulliso axis is another major source of iron ore (Hamrla 1964/1966). There are three sites (Caaggoo and Warra Qaalluu)-two about five kilometers southwest of Gulliso town and the third one (Wiinsaa Qacoo) is in Aira district about eight Kilometers west of Aira town (see Hamrla 1964/1966, Figure 9 for Wiinsaa Qaacoo, and chapter VI).

Ore exploration activities for industrial purpose in Wollega seemed to have been intensified in the post-Second World War Period under the auspices of the Ministry of Mines. There is clear evidence that since the early 1960s a series of attempts to analyze the economic feasibility of iron ore mining (see Hamrla 1964/1966 and field reports in the Ministry) has been carried out by the government particularly in the central parts of Wollega. The study areas of traditional metallurgy in Wollega (Aira-Gulliso) are within the close reach of such ore exploration areas. Various reports at the Ministry of Mines and Energy and other published materials (e.g. Hamrla 1964/1966) indicate that iron ore in Wollega originated from metamorphism. According to Hamrla (1966), metamorphism entailed mineralogical and textural changes of the primary sediments. The paragenesis shows that the magnetite and the quartz waste were re-crystallized. The sulphides might be primary in origin and the presence of small grains of **pyrite** and **chalcopyrite** in the ore could be attributed to the conditions of 'primary sedimentation' rather than having been introduced by thermal solution.

Analysis of southwestern Wollega ore has shown that it is a **magnetite-hematite** ore with very low **Sulphur** and **Phosphorous** content (Hamrla 1966). The percentage of **silica** is low in samples of clean ore but can be higher owing to the variable inclusions of quartz and quartzite mother rock to which the ore is bound. In order to obtain information regarding its formation, polished section of ore has been examined under microscope (Hamrla 1966). The result revealed that it is a porous coarse-grained ore of largely martitized magnetite (replacement of magnetite by hematite due to oxidation or reaction). The description of the nature of the finds of Wollega ore by Hamrla (1964/1966) is as follows: The magnetite is almost completely oxidized to hematite in the form of lamellae (thin plates). The alteration is more pronounced at margins of magnetite crystals and martite extends irregularly along cracks into magnetite.

Limonite replaces martitized magnetite in all directions forming rims and irregular bodies. Quartz grains can be found in the ore sometimes concentrated in bands. There is also exceptional appearance of small roundish inclusions of Pyrite and Chalcopyrite in the middle of unaltered magnetite grains. The following compositions were obtained from chip sample analysis of magnetite-martitic selected clean ore for Yuubdo, Katcho (Wiinsaa Qacoo) and Chago (Caaggoo) localities in southwest Wollega by A. Reagan in 1962 (see Hamrla 1966:447):

Aira town is the administrative capital of Aira district (Figure 10). The present Aira town has evolved from a missionary station, unlike Tulluu Waayyuu of Abee-Dongoro, which has served as a garrison station for the Gojjame and then Shoa armies since the late 19th century (Waqweyya 1988).

Locality	Fe	TiO2	SiO2	P2O5	S	Mn
Yubdo	70.9	Tr.	1.2	0.08	Tr.	0.12
Katcho (W)	70.8	Tr.	3.1	0.03	0.05	0.23
Chago	68.8	Tr.	1.8	0.14	Tr.	0.12

Figure 9 Adapted from Milan Hamrla (1966:447) 'The iron and manganese ore deposits in Ethiopia' in Geologija 9: 439-487.

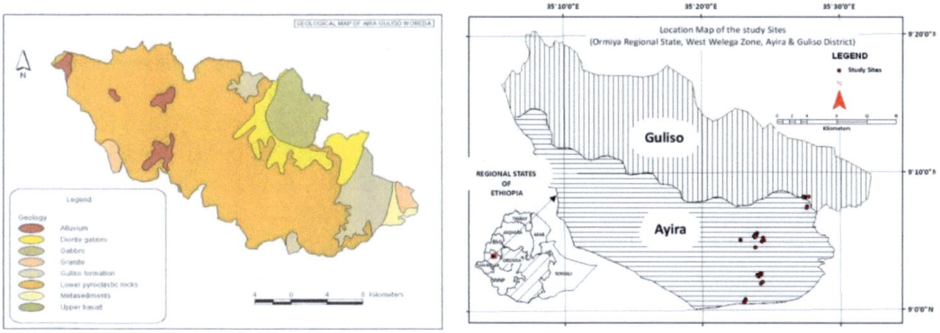

Figure 10 the geological map of Aira district (left) and sites associated with iron production in Aira-Gulliso (right).

Hamrla (1966:50) summarizes the nature of iron ore deposits in Wollega in the following manner: 'the existence of sedimentary-metamorphic iron occurrence in Wollega is... an interesting phenomenon due to the fact that many of the significant world's iron deposits belong to this type, which usually contains high grade iron ore.'

Since the core of iron production (smelting) industrial center is located inside Aira (Ayra) town, one can posit that the reason for its selection by the missionaries must be due to its economic importance. The site might have already attracted substantial density of population before the missionary clinic was established in the first half of the 20th century. According to local sources, the fact that the missionary station gradually moved to the present town (former iron production region) in the 1920s shows that the missionaries have realized that it was where they could deal with more population or the area might have been regarded as a junction for communities of different origin.

Geographically, Aira town is bounded by Jawwee River in the west, by Tulluu Roobaa and Sufar in the east, by Fottoqaa Qorssaa in the north and in the south by Bookkisaa River (Tola 1983). The significance of its location is that in almost every direction, the present Aira town is bounded by small river streams. Most of the former iron smelting sites are in the close reach of the streams, on the eastern and southern parts of the town. One of the river streams to the northeast of the town is the source of high quality clay from whence former smelters extracted clay for tuyères making. Today potters still extract clay along this stream. The town is one of the 17 peasant associations or lower administrative unit (*gandaa*) of the *aanaa*/ district/county. Because of expansion and increase in population, the town's administration has recently divided the town into two administrative parts. According to informants, the dominant lineages in Aira-Gulliso sphere included Jaarsoo, Abayyanii, Laaloo, Gulliso, Bonayya, Alakuu, Baabboo and other minor ones such as Malkee, Guduruu, Innangoo, Innagamdii, Innawaarii, Innabalxii, Innanaanii, etc,.

Aira is one of the areas of southwestern Wollega where missionaries have converted almost all of the communities into Protestant Christianity. Settlers from northern Ethiopia in various towns are however, followers of Orthodox Christianity, the tradition of which they have brought with them from their homelands. There are also a significant number of

Muslims in the towns. As a result, unlike in northeast Wollega where Orthodox Christianity (particularly before 20 years ago) dominated and lived in syncretism with the indigenous belief system of the Oromo, in southwestern Wollega, the Oromo traditional belief system has almost been 'eradicated'. Only vestiges of the belief have survived in former Qaalluu sites. One of these is located east of the town under the name of Warra Abbaa Caffee (Tola 1983). According to Tola (1983), the family of the *Abba* Caffee evolved from ritual importance to chieftaincy by conducting a number of campaigns to subdue the rest of the local clan chiefs partly to control trade including iron. Along a paved road close to the former ritual site are exposed remains (slag in particular) of iron production.

The socio-cultural consequence of the missionary station at Aira was huge. According to informants, the missionary activities (supported with clinics-which today became a highly complex hospital in Aira town as well as missionary schools) did make a number of successful conversions. Nevertheless, the converts at first did not belong to the nobility, such as family members of *Abba* Caffee, who favored to remain with their traditional belief. The nobles and many ordinary families appear to have been unwilling to drop their traditional belief system. Apparently, since Aira and its surrounding was inhabited by huge number of artisans (smelters and smiths), they, particularly the smiths, were the main beneficiaries of the religion and its humanitarian activities. Inhabitants of *gandaa* Jarso Abayyani claim that over 500 sons and daughters of the locality have been educated, found various government and private job opportunities as well as many outside of the country in the western world particularly the USA. Smiths and itinerant smelters exclusively inhabit the Abayyani hamlet of the *ganda* Jarso Abayyani.

The significance of this information is immense. On the one hand, change in the occupational status of the artisans' children created a new opportunity of equal sharing of the resources of their environment. Respect through the success of the offspring has easily been secured. Their socio-economic bases have dramatically widened. This is in contrast to the more conservative areas of the northeast where Orthodox Christianity maintained a strong presence and reinforced the existing situation by introducing into the Oromo region the rigid Amhara hierarchy and strict stigmatization of the artisans (see Burka 2009). There is ample evidence (I have interviewed tanners in Gondar and smiths in Gojjam), among the Amhara population of northern Ethiopia, that artisans were not allowed to serve in the Orthodox Church as members of the clergy or if one changed his social status, he would not be allowed to craft-work, which is regarded as polluting. Despite strict observation of the Orthodox religion, the Amhara secular administration in northeast Wollega approached local society putting its members into one category, that of the conquered and subjects.[23]

23 According to my informant *Obbo* Moosisaa Gaaggaa from the northeast Wollega region of Abee-Dongoro, there were two ways to obtain access to the court of the Amhara during Emperor Menelik: woman and iron. Informants remember, a beautiful noble woman by the name of Goxee in Horro-Guduru. She had no problem in accomplishing her affairs with an Amhara administrator. Similar to Goxee, the person with iron products was more welcomed than the rest of the population. This probably explains how the Amharas upgraded members of the *tumtuu*/smiths to the level of *buuftuu*/ smelters of the mainstream by empowering them with titles such as *balambaras* like *balamb*aras Libidii, father of the well-known master smith in Walagee areas of Abee-Dongoroo.

In this regard, the provision of nominal titles of nobility for both the mainstream chiefly classes and the smiths in Wollega shows the Amhara preference to treat the Oromo communities across the board. However, that does not mean that smiths of southwestern Ethiopia were given equal commensality with the rest of the society due to changes brought through protestant churches. Other historical factors such as brief presence of Italians in the late 1930s to the early 1940s and the Marxist ideology of the Derg regime must also account for the changes. As a result, today, every household in the dominantly craft-working villages (e.g. Abayyanii) own farmlands as well as rear animals. Rules regarding marriage contracts seem to have gradually been relaxed.[24] The question is whether this has affected what the *tumtuu* identity used to connote.

In other words, the issue is whether these changes enable one to discard the *tumtuu* label, which has gone beyond occupational identity to become a social one. This question is rooted in the assessment of settlement pattern of inhabitants of the Jaarsoo Abayyani peasant association. From among the various localities of the peasant association, Abayyanii has the highest concentration of smiths. It does not look accidental. The Abayyanii lineage is generally regarded as being *tumtuu.*[25] Today, in almost all households, smithing is a major livelihood practice (Burka 2008) as well as farming various crops as durra, coffee, and keeping animals.

2.5. OCCUPATIONAL IDENTITY IN WOLLEGA: ITS FLUIDITY AND BOUNDEDNESS

The Oromo society before the 16th century is generally categorized as 'egalitarian' (see Hassen 1990; Leggese 1973; Ta'a 1980; Zewde 1991). This implies that there has not

---

24 *Obbo* Kumarra Sanbatoo is a smith/smelter who organized one of the reconstruction demonstrations at Jarsoo Abayyanii locality. He accompanied me to different villages to conduct interviews. He is a man of about 60 years (or above). One day he took me to an old man known as a storyteller. During our discussion, I wanted to know how the mainstream society looked at the artisans in the past. The story teller, *Obbo* Yaadata Guutaa (over 95 years old) evaded the question many times and when I insisted, he said that the age of the morons has gone. He added that every one of us should live according to the present. I pressed on to know whether he would have accepted marriage demands by the *tumtuu* to one of his daughters or sons. He simply said you are asking me when we were all naive. Later on his grandson, a university graduate, who was in vacation and listening to my interview told me that his grandfather contained the information because of the presence of the *tumtuu, Obbo* Kumarra Sanbatoo. On our way back, *Obbo* Kumarra told me that when he was young he used to visit the man's house because of his beautiful daughters but could not ask one, due to his *tumtuu* status.

25 In one of my assessments of the social status of the *tumtuu*, I have made a list of the household heads, consisting of Innagamdii, Innawangii, Innabalxii, and Abayyanii lineages that inhabit in a hamlet, which is separated from the rest of the lineages by small stream (probably accidental boundary). The list of household heads was 54. According to informants (the villagers themselves), all household heads were *tumtuu*. For them, it did not matter whether the family members of the household forged different occupational identity and dropped smithing. They are still called *tumtuu*. They refer to a school teacher close to the village, who regularly visits his family and still maintained smithing as par time-for them because he knew that he was still a *tumtuu*. Other sons of the former *tumtuu* who never did any smithing practice were categorized, by informants, as being *tumtuu*.

been a marked hierarchical classification in the ways of life of the people (Burka 2006). This may be rooted in the attempt to portray the original inhabitants as 'nomadic hordes' (see Megerssa 1994) who were unleashed to ravage the civilized Abyssinian Empire (e.g. Ullendorff 1960). However, closer examination of the structure of the Oromo society in the present and of the stories of the past provides a remarkable insight into their ranked but not highly elaborated structures. Looking into the status of the smelters, moreover, gives us a view not only that supports the ranking argument above but also the peculiar social organization of the Oromo people (see Triulzi 1996).

Whereas in some of the elaborately structured and divinely sanctioned social hierarchies of the Omotic speaking societies (see D. Todd 1975, 1977, 1978; J. Todd and Charles 1978a), smelters belong to a separate occupational identity, in less centralized ones such as the Tsara people of Oska Dencha, they share similar identity with forgers and potters (Haaland *et al* 2004a, 2004b). The smelter also forged the tools. Inhabitants call the smelter 'the big pot maker' (see Haaland *et al* 2000, 2004a; Haaland 2004a). This name puts the smelter in the rank with the potters, below the smith. A similar social organization is also known in other societies such as the Fur of the Sudan (Haaland 1985). In these societies, smelters are highly stigmatized.

Unlike their Omotic-speaking neighbors of southwestern Ethiopia (see D. Todd 1977, 1978), the Oromo of Wollega (both in the districts of Abe-Dongoro and Aira-Gulliso), give a prestigious position to smelters (Burka 2006, 2009) than the smiths. At least, the smelters claim to possess prestigious position in the society. Unlike their neighbors, the Oromo smelters distinguish themselves from craft workers such as smiths, weavers, potters, and tanners. They trace their lineage from the upper rank of farmers and rulers. In the following section, I will try to show some of the features that enable to recognize ranking among the Oromo society in Wollega.

2.5.1. RECOGNIZING RANKING AMONG THE OROMO OF WOLLEGA

As I have already noted, the Oromo traditionally appear to have been hierarchically structured (cf. the story of the three brothers). There are evidences in their day-to-day relationship that reveal the presence of various forms of hierarchies. The recognition of different statuses in the society is witnessed among both the mainstream and the artisans. Since the Oromo society has established a system by which collective history is stored by certain individuals, the analysis of this data can clearly be biased in favor of the dominant members of the community. Nevertheless, my attempt to look into the structure of the society emanated from lack of clear position of iron smelters (see Figure 10) in the social organization (traditional ranking) of the people of Abee-Dongoro (see Burka 2006). In the end, I also discovered similar type of social organization in Aira-Gulliso. Even if my focus in this section is to examine the social place of iron smelters, the *buuftuu* not the *tumtuu*, among the Oromo of Wollega, this could not be complete without discussing other occupational identities in the society in general.

Despite different views from both sides (members of the mainstream and artisans), the vertical relationship in Oromo society goes beyond the senior/junior relationship (see

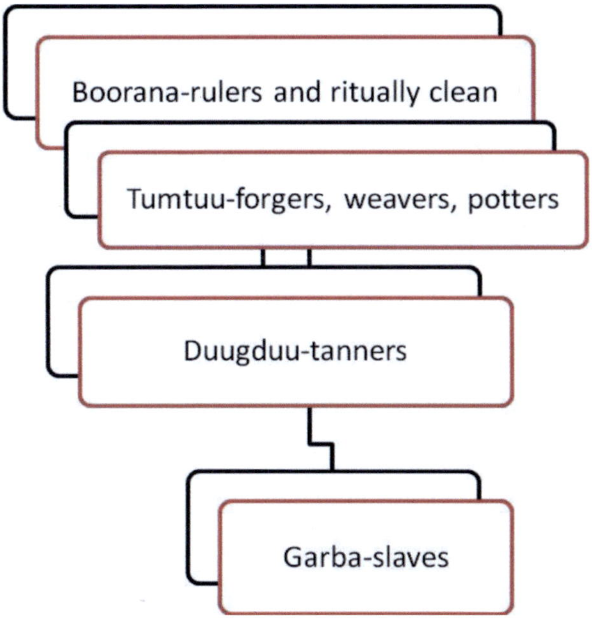

Figure 11 sketch to show the nature of social and occupational hierarchies in Oromo

Figure 11). The senior/ junior relationship only explains the relationship within members of the mainstream[26] Oromo society. In other words, the rest of society including the *tumtuu*, the *duugduu* and the *garbaa* are not party to the senior/junior (*hangafa/quxusu*) categorization.

Hence, the Oromo society has been structured vertically (Figure 10) in that at the top are situated the *boorana/hangafaa*-farmers, rulers and 'citizens'. They regard themselves as the first born of their ancestors. As a result, they claim to have been given seniority and considered themselves as ritually clean. Along the same line, but lower in areas such as rituality are the *gabaroo/Quxusuu*, (the juniors, but who are biologically and socially Oromo). The difference between the *boorana* and the *gabaroo* (see Bartels 1970; Hultin 1984; Triulzi 1996) is that while the former belongs to the first-born son, the latter belongs to the younger in the lineages. In addition to the younger Oromo, the *gabaroo* include those who socially became Oromo through adoption. The *boorana* class claim that they are ritually clean and for this reason belong to the priestly class. I have shown elsewhere that there is lack of consensus on who really belongs to the *boorana* stratum (e.g. Burka 2009).

26 The term mainstream is intended to refer to the non-handcraft population (but not all of them) in the countryside that mainly depended on mixed agriculture (farming and cattle herding). It is to this group that the chiefs, rulers, and owners of the resources of the land belonged. These ones call themselves *boorana*. They claim that they are as well ritually clean and are entitled to bless and curse. This is bestowed upon them by birth. This is to distinguish themselves from those of mixed birth (Oromo and non-Oromo) and from the artisans and the slaves.

The artisans, particularly the *tumtuu* come below the farmers. They include the *tumtuu*[27] (smiths, weavers and potters), the tanners (*faaqii /duugduu*) followed by former (ex) slaves (*garba*) and their descendants at the bottom. The forgers, tanners, and slaves are endogamous caste groups (Bartels 1970). For example, being a forger (*tumtuu*) is ascribed by birth. The son of a *tumtuu* whether or not he takes up the profession of his father is socially a *tumtuu*. To be *tumtuu* becomes a social identity ascribed to him by birth. This status remains whether the son takes up the occupation of his father or changes his status economically. Becoming a *tumtuu* forger implies that he fills a despised and dangerous position as viewed by the mainstream Oromo society, the *boorana* stratum, to which smelters claim to be a part.

However, according to tradition (cf. *chapter VI*), in the past, forgers were not only barred from commensalities such as marriage, but were also not allowed to participate in other social activities such as ritual ceremonies. According to my informants, in the past, tanners were not respected and were made to seat near the entrance of the farmers' houses. An old informant from Aira-Gulliso district (*Obbo* Haayluu Amantee) noted that the mainstream regarded the potters who belonged to the *Waata* as low and polluting. As a result, they had to carry their drinking cups when visiting houses of the mainstream. Even worse, when a *Waata* potter leaves the house of her host, she had to walk out backwards.

The tanners were the most despised but the 'wettest'[28] of all the artisans. Their marginalization was due to their association with the bad smell of hides and skins (Bartels 1970). Consequently, they kept themselves apart from other people (Dafa Jamo 1983 on the Oral law at Odaa Nabee about 1700). Among the Oromo, the saying *gabaan duugduu fi gabaan saree harbuu jala*, (literally means the market of tanners and dogs is under a *harbuu* tree, genus Quercus), expresses their marginalization and location on the edge of the market away from other marketers. Their place was equaled with the status of dogs that live on scraps and refuses and with the tree regarded as lower quality for construction material but productive and fertile.

Contrary to the forgers who are regarded as dry (sterile), however, the tanners are regarded as *jiituu*, (wet). This is also concerning the presence or absence of a good or a bad omen during encounter with a member of the groups. Nevertheless, the general understanding among the Oromo is that the forgers are *gaasha Boorana*-(literally means close to Boorana) and this puts the forgers (including weavers and potters) above the tanners in the social hierarchy. It is noteworthy that the term forgers and tanners represent a feminine gender although the work is in the male domain. The bottom layer, the slaves mainly belong to non-Oromo individuals who became part of the society through slave trade and raids that prevailed until the Italians effectively stopped it during their occupation of Ethiopia.

---

27 The *tumtuu* constitute endogamous handicraft groups made up of black smiths or forgers, weavers and potters. As a result, it is difficult to use as though it represents only one of these groups. However, for convenience sake, I will henceforth use the term *tumtuu* to represent the *tumtuu* to represent only the iron forgers.

28 Wet' is derived from the Oromo term *'jiidha'* and the opposite of 'dry' equivalent in Oromo of *'gogaa'*. 'Wet' is a metaphor referring to fertility. In this regard, the tanners are wetter than the smiths whom the people regard as being dry or sterile polluters (see Pankhurst 2001, for various terms ascribed to artisans for marginalization in different parts of the Ethiopian region).

The other main point to note regarding the social structure of the Wollega Oromo is the absence of iron smelters (*buuftuu*) among various separate identities in the hierarchy (Figure 11). Because the focus of the research is on iron smelting, I had to conduct enquiries into the different socio-cultural and economic relationships within the Oromo society in Wollega. It can be said that it was part of the search for the identity of the smelters. This has emanated not only from absence in the hierarchy but because of former smelters entitlement to the ownership of iron mining sites, its smelting, and transaction. Similarly, this enquiry has developed from insistence by former smelters/*buuftuu* to dissociate themselves from the smiths/*tumtuu* identity. In contrast with the forgers, whom one can distinguish by their settlement patterns, there is no indication of a separate settlement of smelters. The following case study comes from northeast Wollega among the Malkee clan in Abee-Dongoroo district. Since the same tradition is working for the southwestern Wollega smelters (where members of the mainstream explain that 'people like us' smelted iron and that they have immigrated from eat Wollega), the data from the northeast represents the explanation for the status of smelters/*buuftuu* in both study areas.

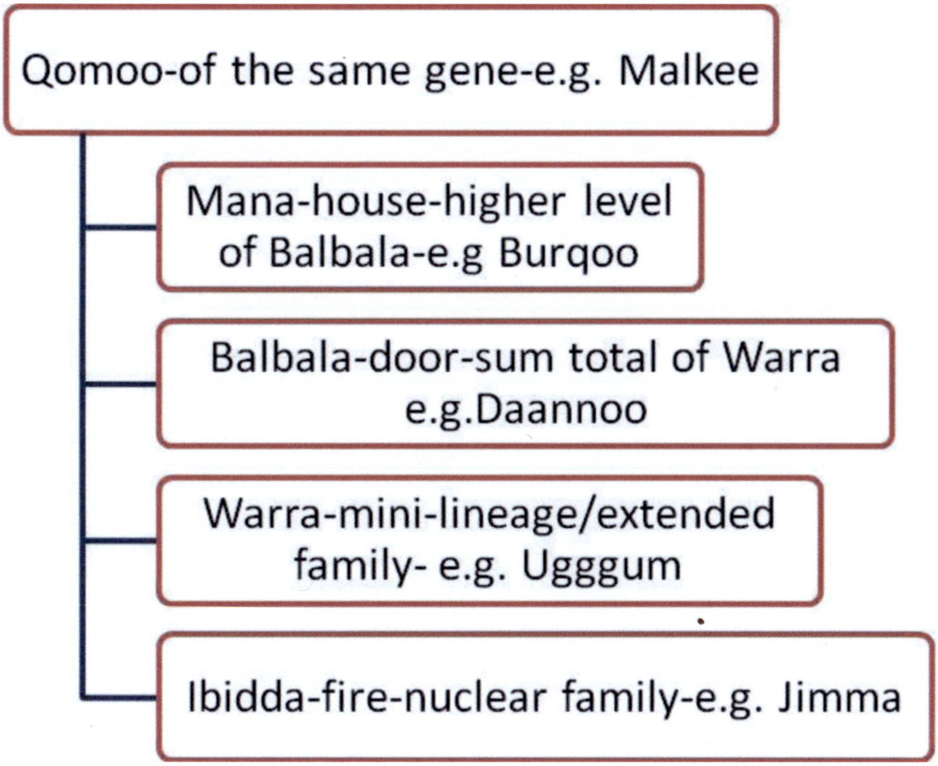

Figure 12 Lineage relationship in Oromo of northeastern Wollega, a model developed from the genealogy and senior-junior relationship of the Malkee clan, the iron smelters

The first attempt I made was to look for the presence or absence of separate identity of smelters within the mainstream society. One method I employed to find this identity was to look into the marriage relations of smelters. I closely examined whether they were endogamous or exogamous. I also looked into the traditional Oromo land tenure system in the past. In this case, I concentrated on what level of ranks the clan/*qomoo* owned or controlled and redistributed land or any common property among the Oromo as well as locating the center where power concentrates most.

This endeavor directly led me to examine further, what is known as the senior/junior (*hangafa/ eegoo or quxusu*) relationship and its roles among the people. I found out, that though the Oromo society from the outside looks likes an egalitarian system, it is internally stratified. As a result, the system puts one individual or group at the top of the other (Figures 11&12). It is automatic to become senior for the first born in the family. Nevertheless, it also signifies ritual status making the senior a ritually clean representative of the group. In other words, being senior means being a *boorana* and being *boorana* means being ritually clean and a mediator between the people and Waaqaa, the Oromo supreme God.

To elaborate this further, it was essential to observe the Oromo social structure below the clan and examine at what level of the clan the function of seniority becomes evident. Seniority is clear at clan level, however. On the other hand, there are clans that are senior than others in that not all clans enjoy equal status when it comes to such roles as blessings

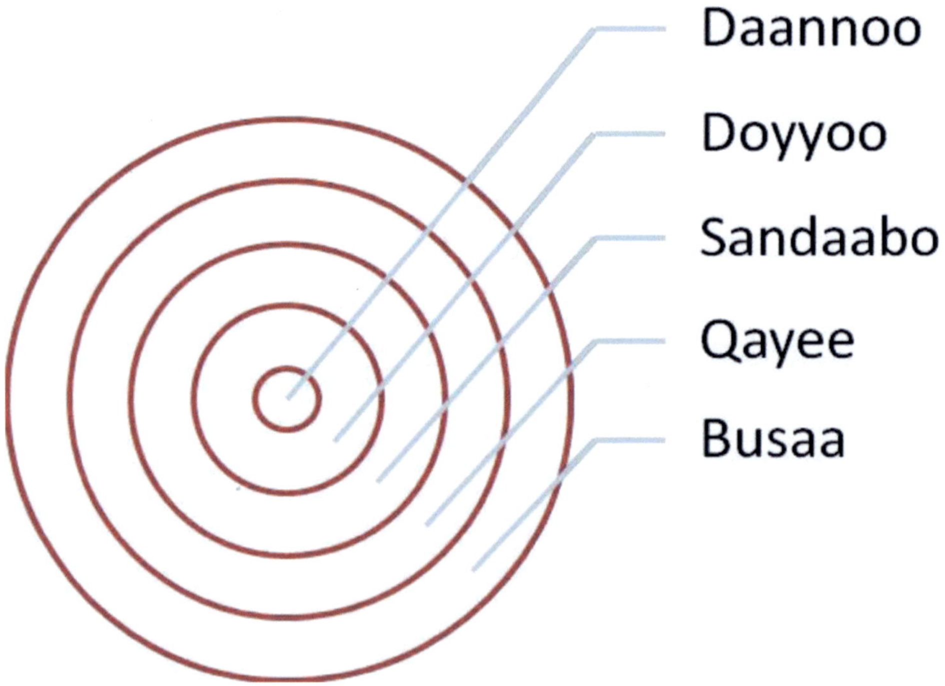

Figure 13 Senior and Junior Oromo settlement pattern (centre-periphery)

and sacrifices. However, since our interest is to reexamine the structure at clan level, we should not go further to discuss clan-to-clan comparisons. Seniority becomes clearer when it is seen at lower level of the clan called *balbala*. A clan is made up of the sum total of *balbala* e.g. Malkee is composed of five *balbala* (see Figure 13). Among these *balbala* one is senior (*hangafa*). The senior *balbala* possesses right of leadership (political, military, and cultural such as spiritual) and exercises control over property (economy) such as land inhabited by members of the clan. The *hangafa* are the Boorana-people of the morning-meaning the beginning. The *hangafa balbala* and the *hangafa warra* distribute wealth to members of the clan (see Figure 12 & 13).

Through the rules set by the *hangafa halbala* and *warra* members of the junior *balbala* and *warra* own property, occupy land, and use it. In return, the juniors were expected to live as tribute payers to the seniors or move further away from the control of the seniors. The Malkee clan is made up of five *balbala* (Figure 13), namely Daannoo (*Hangafa*), Doyyoo, Sandaaboo, Qayee and Busaa. These five *balbalaa* lived side-by-side occupying different parts of Walagee with the seniors taking the central position and the juniors living in flanks (Figure 13). The Daannoo, the senior *balbala,* were inhabitants of areas around ore mining site in Walagee. It was probably coincidental that the *hangafaa balbalaa* occupied a place of an important resource, the iron ore, or Walagee.

There are four *warra* under *balbala* Daannoo, namely *warra* Dagaa, *warra* Looyyaa, *warra* Ayyaanaa and *warra* Uggum. The *warra* Uggum Aagaa has run the control of power and resources in Walagee. *Warra* Uggum belongs to *hangafa warra* under Daannoo, which in turn is *hangafa balabala* of *Qomoo* Malkee. As a result, *warra* Uggum and other relatives in Daannoo have handled distribution of wealth in Walagee (Figure 13). Next to *warra* Uggum, warra *Balaambaras* Horroo Jimmaa were in charge of the redistribution of resources such as iron ore.

Generally, other members of *balbala* Daannoo had natural right to have access to the ore as well as other properties through permission from the *hangafa warra*. Other members of the clan obtainedthe ore by giving necessary services (e.g. labor for extraction of the ore,

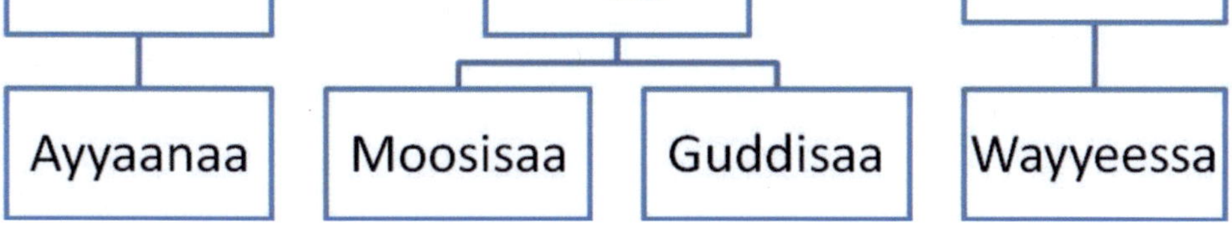

Figure 14 elaborating the relationship between seniority and power inside closely related family

working on bellows owned by the *hangafa balbala* and *warra* members) and tributes to the elders. In this classification, the handicraft groups were excluded. The next question is who smelted the ore? Did the smiths, the endogamous caste, do it? On the other hand, did the *boorana,* the mainstream, carry it out and still claim that they were chiefly and priestly class? How did members of the Malkee lineage appear to other lineage members of the mainstream in Abee-Dongoroo?

Outsider clans in Walagee and Abee-Dongoroo generally regard the Malkee clan as *buuftuu* (a feminine), literally 'blower'. Hence, the smelting place (hut) is *godoo buufaa*, blowing hut. According to informants, almost every member of the clan in one way or another participated in the mining and smelting activities in the past. The ones who played leadership role had their own smelting huts. It is interesting to note also that the wives of immediate relatives of Uggum had their own blowing hut to run. According to informants from family members, the wives made their dependent relatives to work as smelters. The dependents or any member of Malkee or Daannoo or members of junior *balbalaa* who could not afford to organize their own blowing hut worked as *baajjattuu*, fokaachii, or *rabbacii.*[29]

Let me give a more concrete example (Figure 14), which shows that in spite of their control over the ore or quarrying pit, some members of the senior class worked as master smelters or participated in smelting activity or mining. The following lists of individuals belong to Jimmaa in the Daannoo lineage. First are *Obbo* Moosisaa and Guddisaa Gaaggaa Jimmaa. These two were my informants and aged over 90 years, the former died in 2008. They were only two surviving grandsons of Jimma. The second group is *Obbo* Wayyeessaa Bokoree Jimmaa, and *Obbo* Ayyaanaa Horroo Jimmaa. Out of these three lineages of sub *warra* Jimmaa, Ayyaanaa Horroo belonged to the senior. In other words, Horroo was the first son of Jimma (see Figure 14). In spite of this, informants claimed that all members of the sub-lineage had equal right of access to the ore. From this group, Wayyeessaa Bokoree is known to have been the best master smelter remembered by everyone in Walagee. However, Moosisaa and his younger brother Guddisaa also participated in both quarrying and smelting whereas Ayyaanaa Horroo only oversaw the activities.

In order to increase our understanding of iron smelters in Walagee and their position in the society, it was necessary to look into their marriage relationship. Outsiders regard members of this *qomoo* as *buuftuu* or blowers. Because of this, they were given a nickname- *Malkee Wayyaa Xurii* (literally 'Malkee of dirty clothes'- because of the soot from charcoal and smelting). The question is whether members of this clan were accepted in marriage or not outside their occupational identity. This forces us to deepen our scrutiny. From the derogatory term, *wayyaa xurii*, ascribed to the smelters, it could be concluded that no other group outside their identity, particularly the farmers, would enter into marriage contract with them. However, according to informants and marriage histories, there is no single case where members of this group (in the Figure 14) were prohibited from entering marriage relations with notable *qomoo* leaders in Abee-Dongoroo and beyond. Five of my existing

29 It may be confusing to hear three words from different informants. *Baajjattuu, fokaachii,* and *rabbaccii* stand for similar meaning to mean those who provided free labor in return for free food and bloom fragments at smelting sites in the past.

informants from members of the smelters (Wadeessa Danjee) aforementioned traced their exogamous marriage relations as opposed to only one exogamous marriage from the *tumtuu* hamlet of Masoo and one from Somboo. Therefore, the term *malkee wayyaa xurii* seems to have been superficial. Even more, as members of the smelters claimed, members of the mainstream society carried out iron smelting.

Consequently, the marriage evidence, we have, can serve as a basis to conclude that among the Oromo of Wollega, iron smelters belonged to the top social class as farmers and rulers and enjoyed prestige in the society. As I have attempted to indicate at the beginning of this chapter, among the Oromo, iron smelters belong to the *boorana* (rulers, farmers, citizens) upper hierarchy followed by forgers/smiths, tanners and finally slaves at the bottom of the socio-economic hierarchy. Such categorization may not be, however, representative of the whole Oromo who is widely distributed across the Ethiopian region and the Horn and exposed to various dynamics. Such categorization of the smelters with the chiefly class while leaving the *tumtuu* at bottom is not acceptable to members of the *tumtuu* (sees Burka 2009; also cf. chapter VI).

Generally, however, the Oromo are divided into five moieties or halves. These are the Sabboo and the Gonna, the Macha and the Tuulama, the Raayyaa and the Asabuu, the Siikoo and the Mandoo, and the Ituu and the Humbaana (see Megerssa 1994; Kassam 1999). The division of these descendants follows the occupation of different regions. In this work, I focus on the Macha group in Wollega of the Maccha and Tuulama moiety. Accordingly, the region of Macca and Tuulama included most of the present Shawaa, Jimmaa, Iluu-Abbaabooraa, Wollega, as well as areas north of the Blue Nile in Gojjam. Even within the Maccha Oromo of Wollega, it must be noted that there exists variability and diversity across space and time due to various historical circumstances including exposure to different neighbors. The following chapter takes data from two iron smelting sites in Wollega with the aim of situating it in theoretical and object of study of the archaeology of iron production, known as the *chaîne opératoire* framework.

CHAPTER THREE

3.0. VARIABLES IN IRON METALLURGY: UNLEASHING THE *CHAÎNE OPÉRATOIRE* FOR THE ETHNOARCHAEOMETALLURGY IN ETHIOPIA (WOLLEGA)

3.1. INTRODUCTION

> *....In accounts of Himalayan communities, every interview began with an obligatory cup of tea embellished by a layer of sour yak butter floating on top. As the story of each interview approaches, the reader can virtually taste the rancid butter and experience its slimy mouth feel, but one could not feel the cup in which it is served... (Gloria London 2000:2)*

The archaeology of iron production through bloomery smelting is a complicated task mainly due to the difficulty of observing what exactly takes place inside the furnace and prone to changes in response to variables. Also problematic is in the absence of advanced methodology, such as in Ethiopia, to conduct analysis of finds in order to provide plausible interpretation of the technological achievement attained that begotten the objects. The technological complication of metallurgy and lack of developed scientific facilities require various approaches for acquiring knowledge of the process that brought about the material culture. This is because the aim of an archaeologist should go beyond analysis of the end-products in the archaeological record i.e. dealing with the dynamics as well as the statics.

The archaeological study of metallurgy in Africa has long diverged from the approach developed by its western counter part in that it focused on the ritual and cultural components than the physical aspects of the craft (cf. *Chapter I*). I propose two possible factors for such a distinct departure. The major one is the fact that it is in Africa that iron metallurgy, a traditional iron production practice, has continued until the recent past. This provides ethnographic data as not only once observed and recorded when it was active, but also that there is a possibility of reconstructing the activities, which stopped only about half a century ago. The African traditional smelters' approach, which made use of a number of ritual interventions, during the production activities, might have influenced the nature of the technology's study in Africa.

The second one is that scholars who did not have ample technical expertise of the physical sciences conducted the earliest archaeometallurgical studies in Africa. As a result, they did not go for the search of necessary laboratory facilities to explicate their observations. In addition, those who engaged themselves in the study of African traditional metallurgy probably believed that they could conduct sufficient study without the use of any scientific facilities. It implies that African technology was either to be explained by methods established elsewhere outside the continent or not considered to require much scientific tools to analyze (see Mapunda 2000). Apparently, the term bloomery smelting technology is sought in African iron metallurgy elsewhere in the continent (see Schmidt 2001).

In light of the first point, it is necessary to underscore that the archaeology of iron production is best understood, when the archaeologist is able to observe the whole processes that begotten the objects, not just the end-products. Thus, in the absence of advanced scientific tools to analyze the few surviving archaeometallurgical debris and because of available ethnographic data, the study of iron production in Africa follows the procedure that involves the observation and documentation of the production trajectory focusing much on its ritual aspects (see Childs and Killick 1993; Killick 2004; Van der Merwe and Avery 1987).

In this case, the conviction is that the concept opens up vistas to access to information on the various physical and non-physical aspects of the production activities. The difficulty in accessing information through the study of archaeometallurgical debris such as slag mounds, tuyère fragments, collapsed furnace walls, remains of furnace foundations etc., is resolved through an ethnographical study of metallurgical activities that employs the *chaîne opératoire* framework. It becomes more enlightening when, for instance, the data allows using a direct historical approach, the situation under which the present communities of the study area could provide 'direct' link to the archaeometallurgical remains (*e.g. cf. Chapter V*). It is, therefore, an excellent opportunity for archaeologists to observe how humans negotiated with the material world in their various ways of adapting to their respective environments. In other words, the ethnographic study of metallurgy provides important opportunity for the archaeologist to grasp a picture of the negotiations between the human and the non-human world. Accordingly, such approaches as the *chaîne opératoire* enable to amply investigating the various constraints either natural or cultural.

This chapter discusses the different technological and socio-cultural variables and the constant behaviors involved at different stages of the 'indigenous traditional' iron metallurgy (*cf. chapter I*). This chapter elucidates the significance of the concept of ***chaîne opératoire*** in designating the procedures undertaken in the ethnoarchaeology of iron production processes and activities. Ethnoarchaeology is all about the study of material culture in their ethnographic setting. In other words, it is the study of the objects throughout operational sequences involving their production, use, and discard. In the following section, I will discuss the concept of the ***chaîne opératoire*** in light of perspectives developed by different scholars to underscore its applicability in various fields of study. Then, I bring in the phases in the ethnoarchaeology of iron metallurgy in Wollega with the aim of elucidating the applicability of the ***chaîne opératoire.***

Scholars agree that the concept of ***chaîne opératoire*** is associated with and credited to Marcel Mauss (1872-1950) and Andre Leroi-Gourhan (1911-1986) (see Martinon-Torres, 2002; Schlanger 1994). According to them, (see e.g. Martinon-Torres, 2002) the ***chaîne opératoire*** as a concept or object of study is so flexible that it has come to be applied to various fields of study (see Schlanger 1994). It is because the ***chaîne opératoire*** can be 'freed and spread' and such malleability explains why various scholars began to utilize it in various fields of study (Martinon-Torres, 2002). Even then, the ***chaîne opératoire*** both as a concept and as object of study (Barndon 2004; Chirikure 2005; Martinon-Torres 2002; Schlanger 1994) is commonly linked to the study of prehistoric stone tools manufacture.

Nevertheless, the ***chaîne opératoire*** is an appropriate object and / or approach for the study of any technological processes engaged with the transformation of things (see

Barndon 2004; Chirikure 2005; Thornton and Giardino 2008). Indeed, my attempt in this chapter is to show how this approach is adapted to local situations for providing theoretical basis in explaining the subject under discussion-iron metallurgy. Since the archaeology of iron production is an endeavor to replicate the stages involved in the transformation of materials, the ***chaîne opératoire*** concept provides an appropriate ground for the ethnoarchaeological study of metallurgy such as in Wollega. In the following section, therefore, I first present the concept as defined by one of its early theorists and then show its applicability to the study of traditional iron production that involves the transformation of things through various interrelated stages.

Andre Leroi-Gourhan (Cresswell 1990:46) defines the ***chaîne opératoire*** as 'a series of technological operations which transforms a raw material into a useable product' (see also Martinon-Torres 2002: 31). The significance of the concept according to Cresswell (see Martinon-Torres, 2002) who draws from Leroi-Gourhan is twofold. The first is its behavior to enable to show the trajectories in successive operational sequences, and the second is its peculiarity for bridging the human and the natural world rather than viewing them separate. With the second regard, the archaeology of iron production, from my standpoint, best represents the inseparability of the human and the natural forces. This is particularly important where the discussion addressing the archaeology of iron production is among traditional societies such as in Africa. In such a society, the boundary between nature and culture is completely blurred.

In order to distinguish the distinctive place the ***chaîne opératoire*** approach occupies in the study of technology, it is worth mentioning the shortcomings of other theoretical frameworks such as culture-historic and the so-called the New Archaeology in the study of technological achievements. Whereas the culture-historic approach propagated a diffusion/ migration and/ or evolutionary model in explaining culture change, the New Archaeology linked environment and technology to explain human adaptation to nature and the change because of this (Martinon-Torres 2002). In both, as further explicated below, the focus is on the being than on the becoming of things.

According to Martinon-Torres, both approaches focus on the outcomes, lacking the peculiar qualities of the ***chaîne opératoire*** to address the specific human activities involved in the becoming of the material culture. Both previous approaches are concerned with the search for universalities in the laws that govern the creation of material culture. They lack explanation for particularities and the nature of their becoming at various stages of production, use, re-use, and discard. It is, however, in the becoming of material culture that we are able to observe and distinguish between the roles played by humans and the environment to constrain or facilitate technological activities. On the other hand, the previous approaches concentrate on the physical aspects of things. Since they focus on products at their final level, they have no means of explaining the non-physical aspects of things.

In contrast, the ***chaîne opératoire*** allows one to observe the non-physical aspects surrounding the physical aspects of things during production processes. The ***chaîne opératoire*** approach enables us to deal with such queries as, for instance, whether it is possible to infer the actors' choices or decisions in bringing the matter throughout the production procedures. It also allows dealing with the flexibility attained within the constraining environment or culture. In addition, the approach enables to distinguish the real factors affecting production activities. These include whether natural environment determines/facilitates production of things or the technology applied is the major factor or the choice of humans are the most important ones. In other words, it enables to deal with issues as 'environment as a determining factor' in cultural development or whether it is possible to manipulate it by cultural aspects of different societies.

Scholars further posit that (see Martinon-Torres, 2002:31; also Barndon 2004; Schlanger 1994) '... the concept of ***chaîne opératoire*** puts emphasis on every stage of the tools production, particularly the supply and properties of raw materials. The object of the study becomes thus 'longer': research focuses not on a single item but on a sequence....' Nathan Schlanger (1994:144) puts the concept as '...a practical, operational and coherent means with which to pin-point the part of the artificial, the interplay between fixed and flexible'. We can draw two important concepts from what scholars posited above, for the purpose of this chapter. These include 'stage and sequence'. I believe the two concepts refer to the significance of meaning of things in chains or in their relations back and forth. Stages and sequences show things in length and enable us to travel through the trajectory of production life cycle. Human decisions are held in stages and stages are driven in sequence. These stages enable variables to work together so that the existence of the previous is essential in itself and for the realization of the previous and the activities to follow.

This entails that the chains fasten together the human (social-labor, finance, ritual, power) and the natural (environmental-iron ore, clay, hardwood, and charcoal) behaviors constituted by material culture of production. The stages that involve different levels of production processes are also chained, working not in isolation but in association. In iron production, the stages include-mining, hard wood selecting and charcoal making, clay extraction and furnace building and tuyères making, house construction, smelting, smithing, trading, use, repair, and discard. The decisions under previous stage condition the nature of actions to be taken in the subsequent stage(s). Previous stage is not an end in itself. It is part of the horizontal and vertical processes consisting of factors causing the next action. It is not final at all. Revisiting the precious action is part of the process. In this case, the subsequent stage is inseparable from the former. Therefore, one stage is not complete by itself; it is incomplete on its own but still useful and integrates different aspects vertically and horizontally in 'width and length'. It chains together both human and environmental factors of production (see Figures 15&16).

The philosophy behind the 'width and length' can be conceptualized, for instance, as what is the significance of collecting iron ore if it does not transform into bloom in one of subsequent stages? What is the significance of building a forced draught furnace with one-meter radius and two meters tall, if there is no enough charcoal to reduce the ore or if there is lack of labor to work on bellows? What is the importance of making a one-meter-and-half long tuyère if the furnace is not tall enough to accommodate it? Therefore, since the concept provides us the lights in sequence, it markedly differs from end-product material-culture-approach as studying 'types' to determine culture, or where environment is seen as a determining factor for causing changes in a society.

The concept of ***chaîne opératoire*** is well founded in the study of stone making technology (Barndon 2004; Schlanger 1994). Notwithstanding this, we find that the framework is fitting to the discussions of the archaeology of iron production (e.g. Barndon 2004; Chirikure 2005). The ethnoarchaeometallurgical evidence in the study area (Burka 2006, 2008) has inspired the application of the concept. This is because it strives to answer a number of questions regarding iron production activities at every stage. These questions range from technological design to production activities, to use and repair and to the discard of the materials as well as to the cultural thoughts embedded in each level of the activities. It is not only the physical products of activities that the concept makes possible to attain but also the various human behaviors embedded in the process of production. In other words, the concept of ***chaîne opératoire*** serves as an interface between the physical and non-physical aspects of production activities. In light of the core of the study, ethnoarchaeometallurgy, I first establish its theoretical basis through the ***chaîne opératoire*** framework upon which I build *chapter V* that provides empirical data through reconstruction and re-enactment of iron production activities in Wollega.

There are at least six variables important in the production procedure of iron involving operational sequences (Figure 14). This chapter mainly deals with the technological aspects of the iron metallurgy. Nevertheless, since the non-physical aspects (social, cultural) of iron production are virtually inseparable from the physical aspects (technological), there are many instances where the two aspects are concurrently elaborated. Moreover, these take place at different levels of the production and use sequences. The choices, decisions, constraints, and flexibility within these realms give us opportunity to observe the human-material interactions. Guided by the light of the ***chaîne opératoire*** concept, I have categorized the 'operational sequences' in traditional iron production activities into six phases. The six phases are: 1) the procurement phase, 2) preparation phase, 3) charging phase, 4) the smelter-smith-trader communication phase, 5) the smithing phase, and 6) the dispatch phase. The discard procedure is only addressed in connection with refinements at different levels as well as the records of survey evidence at various past production sites (cf. *chapter IV*). Given the time of study, however, it is farfetched to acquire data on the after use periods. Each of the phases is unique but the meaning is incomplete without the rest of the other phases.

The ***chaîne opératoire*** approach enables the research to reveal the basic set of operations in the iron smelting procedures (e.g. Childs 1991). In this case, it allows to record, as for instance, the procedures involved in locating smelting sites situated *vis-à-vis* the village and variables of production. In addition, matters related to the degree of flexibility in human decisions include the selection and processing of materials or construction of smelting houses, building of furnaces, making of tuyères, and bellows, and charcoal making. Furthermore, it helps to analyze whether these inputs affect where to locate the activities and the amount of labor needed for realization of the work.

There are other areas where the concept of the ***chaîne opératoire*** becomes crucial research tool. This includes such queries as whether the socio-cultural (ritual) aspects are observable in all stages of production activities. To be more precise, it allows pinpointing at what stage is rituality critically needed during production processes. Is it possible to discern the rituality of the orientation of the smelting furnace, for instance? Were producers

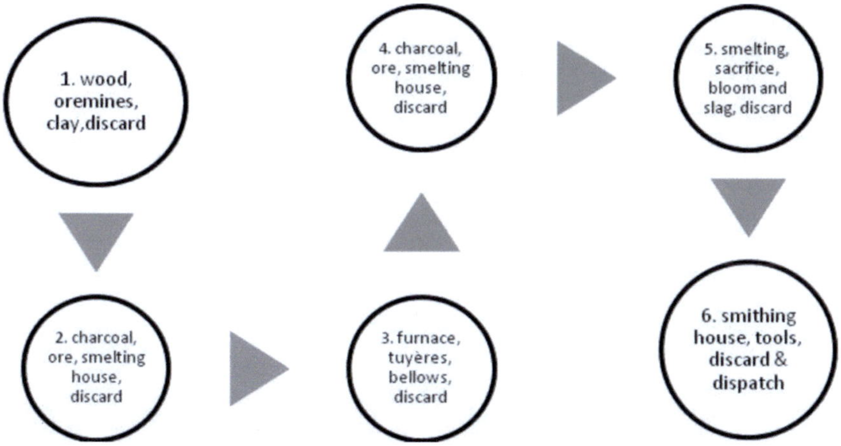

Figure 15.The operational processes involved in iron smelting and smithing

are left with choices on the kinds of objects being distributed in the community? Were the various meanings of objects are coined during their production or use? It is also important to employ the concept in order to identify the connectionbetween rituals during production with the myth of the origin of the material or the artisans. The following are descriptions of the various stages/phases through which people and objects pass during iron production activities (see Figure 15&16).

Phase I: The procurement phase

This is the natural environment stage. It is the stage where humans and the environment interact during the attempts of humans to procure their needs. With regard to iron metallurgy, it is the stage where artisans intervene with the environment to obtain iron ore, charcoal, construction materials, and clay. The stage allows observing the significance of human decisions and the role of the environment to curb or influence those decisions. At this stage, it is possible to examine critically why humans decide to act certain ways, while leaving aside what would look like more logical to us than those they chose.

Phase II: Preparation phase

This is the stage where the processing of the natural objects for production starts. However, this does not rule out the presence of human-environmental relationship preceding the procurement of resources from nature's stock. The second stage is, however, where natural resources begin to transform to cultural ones despite the fact that the boundary between the two is not as clear as it might seem. These include the preparation of smelting site and facilities, for instance, the smelting house, furnace, *tuyères,* and bellows, which were extracted during the previous phase. There is, however, the length between the previous and latter phases where transformations from natural to cultural aspects are taking place. Here, therefore, it is not only the objects that interact with one another but also the various people whose specialty differ on the becoming of the objects including ore miners and processors, charcoal makers, clay extractors and processors for furnace and tuyères, as well as those in charge of the resources. The observer begins to see how people deal with one another in the medium of the objects that they procured from nature.

Phase III: The charging phase

Between the second and the third phases, processes continued to transform things and people. This finally culminates in a setting where the still not fully transformed natural objects (e.g. iron ore, the tuyères) undergo further changes. At the third stage, further material and social ingredients continue to interact and undergo changes. It is one of the critical stages where rituality is saliently embodied. Smelting-the process of cooking the raw- takes place here. At this stage, new ingredients other than those in previous chains appear while some of the previous components are dropped out. For instance, some new people come to the new scene while some others leave the space. In the aftermath of it, a transitional zone, it is possible to observe the engagements between producers and users including bellows chargers, the *baajjattu*, people who provide free service for scraps, the owners of the smelting industry, and those who transform the bloom to other cultural forms, the smiths (see Burka 2006, 2008). The negotiation between materials and people during this transitional period would ultimately inform the ensuing of the next phase.

Phase IV: The smelter-smith-trader communication phase

At this stage, the chain of process helps to observe the occurrence of a number of technological and cultural behaviors. It is the stage where the first discard of the cultural product, slag begins. It is not, however, fully a cultural zone. There are still some natural objects employed into this phase (plant climbers, stone anvils, and hammers). Other activities include refinement and discard of the unwanted materials. It is another transitional zone between the smelting and the smithing sites. Simultaneously, smelters, smiths, beggars, old ages, widows, and landowners can converge and negotiate over the product. Reappraisal is part of the process under this phase.

Phase V: The smithing phase

Once the bloom, the raw iron, is at the disposal of the smiths, they are the ones to transform the thing into desirable cultural products. The *buuftuu* has no physical access to this stage. It is the domain of the *tumtuu*. The *buuftuu* does not forge things. Conversely, the *tumtuu* does not have access to the knowledge for transforming the ore into the bloom. However, their connection and negotiation continues to maintain in the medium of iron products. There is a reciprocal relationship. However, like in all stages, many other elements also

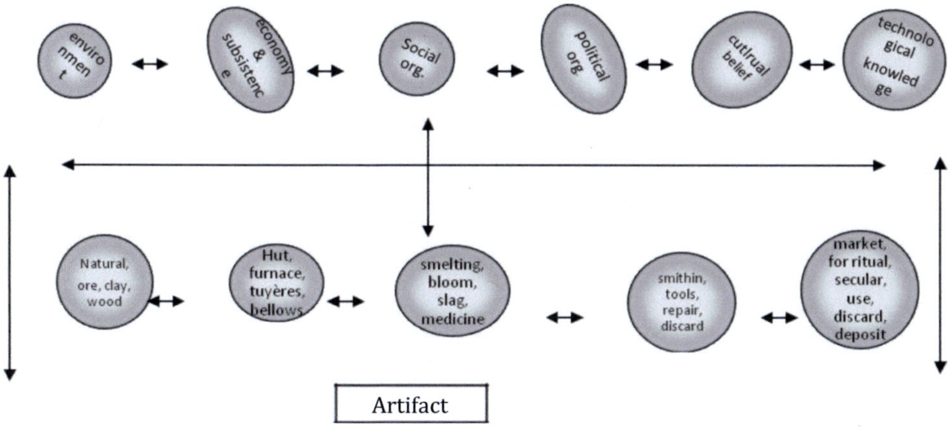

Figure 16 chart for the chain of operation sequences in traditional iron smelting

take part over the negotiation of the forging and the ways of appropriation of the cultural product. The *buuftuu* reprises his technological achievement based on the feedbacks from his customers such as the *tumtuu*. This implies that the *tumtuu* is not just mere dependent on the *buuftuu* (see Figure 15). At this stage, with his special skill,the *tumtuu*, conducts further serious refinement of the unwanted materials and creates new things. During the process of refinement and transformation of the bloom into desirable objects, many unwanted things go to the category of the discard while material, which served various purposes are brought for repair or for change of their use.

Phase VI: The dispatch phase

At the smithing stage, the *tumtuu* forges various objects for various purposes. His products can be dictated by the nature of the demand in the society. There is, however, a room for smith's choice. The tools could range between small utilitarian objects to big weapons. This determines the destination of the objects after the smith fashioned them into various designs. However, the same type of object can contain different meanings due to use value ascribed to it (see Burka 2009). The *tumtuu* might receive orders to forge things either for secular or ritual purposes. As a result, the role places him in ambiguous position as well as the objects. Even then, intermediaries who dispatch the products into the markets can influence the decision of the *tumtuu* in forging certain tools. In this way, the *tumtuu* has the choice of selling his products either on the spot at the forging site or transport them to market places where various kinds of people gather (see Figure 15). It is after this that their use is mainly determined i.e. either for secular or ritual purposes although selected incidents of special orders can happen. The way objects are used after they left the market determines how they are later on disposed of. The use, the disposal stage, and the post disposal processes are at far level away from their production phase (see Figure 15). The chart (Figure 16) can further elucidate how these activities are coordinated and chained together to provide symphony.

3.2. TECHNOLOGICAL AND SOCIO-CULTURAL VARIABLES IN BLOOMERY IRON SMELTING

3.2.1. CHARCOAL

Necessity is the mother of invention. As Man has been faced with sever survival challenges, certain extraordinary individuals have seen, often in a brilliant flash of inspiration, how to address the challenge of Need, by applying the forces, potentialities, and affordances of Nature to the fabrication of tools and material artifacts.

(Adapted from Bryan Pfaffenberger. 1992:494)

The question, however, is can the standard view of technological innovation (see Pfaffenberger, 1992) purely serve as explanatory tool for traditional iron metallurgy? In the first stage, the operating sequences of iron production lay a number of activities into effect. It is the stage where human direct contact with what is regarded natural begins to create a modified environment. According to the standard view, the trees were there to be used for various purposes. It was up to the humans to select from the jungle a tree species believed to satisfy a particular purpose. It probably needs an ingenious person among members of the society to select one species over another or achieve a desirable result through trial-and-error. It is true that charcoal makers are well acquainted with the type of trees needed for different uses. The way they produce charcoal, however, varies according to the experiences they had accumulated from their forerunners or their respective cultures. Ethnographic evidence from Wollega is informative regarding human choices and adaptation to the requirements of one's respective environment. It is not as simple as the standard view of technology.

One of the objectives of this chapter is to demonstrate the change and continuity in different activities constituting iron production through ethnoarchaeological setting. In this case, it became essential to find if charcoal production activities elucidate such an enquiry. Cases from east and west Wollega administrative zones provide us with such a possibility. The study clarified that there is no uniformity and stagnancy or continuity as to how charcoal is

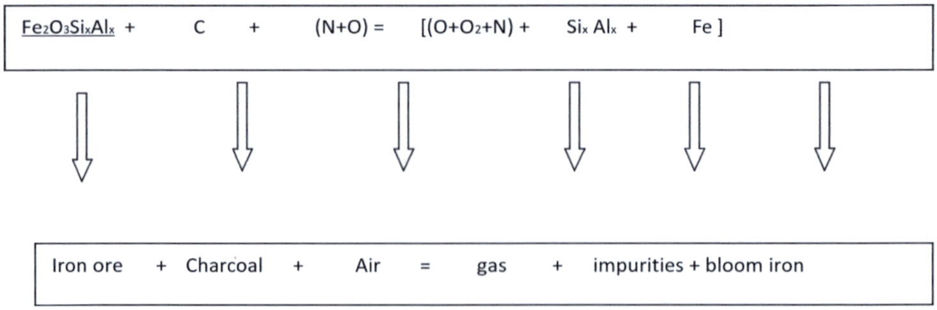

Figure 17 Chemical reactions in the bloomery furnace of traditional iron smelting

prepared either for smelting or for smithing (cf. *chapter IV*) purposes. This is the case also with the choices made on the tree species used for charcoal of these different purposes. Nevertheless, whether smelters were aware of the fact that charcoal prepared in pits, which is rich in carbon (see Cline 1937; Chirikure 2005; David and Kramer 2001; Miller and Van der Merwe 1994) is productive than the one prepared in open space is yet to be resolved. The question that remains is to be addressed is whether such a decision was just cultural (see Figure 17) or dictated by necessity.

During my preliminary research on iron smelting practices in northeast Wollega in 2005 (Burka 2006), I have documented the explanation and criteria smelters and smiths set in distinguishing and choosing a certain tree species over others for smelting and smithing, (e.g. *Hambalta* and *Baddessa* respectively). I indicated that smelters/ smiths did not have any clue of its reduction due to its carbon content, but offered a different explanation. For them, charcoal makers select certain tree species for smelting/smithing because they think that the 'saliva' (vapor) produced by the charcoal during reduction/forging enabled iron fragments to coalesce into a bloom or fragments of the bloom into forgeable material. They argue that the same type of tree is not used for smithing as the one for smelting, because if they did, the iron melts down instead of smelting. In other words, the charcoal from a certain hardwood, which reduces the natural ore into metallic iron, is believed to liquidate (melt) the bloom if it is used at forging station. The descendants or close lineage of smelters and smiths in southwest Wollega, however, do not make such a serious distinction.

In northeast Wollega, charcoal makers prepare it in two different ways: in the pit and in heaps (Burka 2006). It is difficult to trace the origin of both practices. The ethnographic record shows that charcoal for smelting was prepared in the pit and for smithing, in heaps. During the procedure, the half burned wood is quenched with water. Leaves are heaped on top of the half-burnt charcoal and they dump it under a soil. Then it gradually burns for a number of days. This was in contrast with charcoal for smithing, which they prepared it in an open space. Their explanation for such categorization has to do with their belief that charcoal prepared in a pit contains high moisture to coalesce the iron particles into solid mass. One common practice between charcoal for smelting and smithing is quenching with water. The smearing of water over the charcoal even during the process of smelting appears to reflect an aspect of technological input played by other variables in the smelting activities.

The second study area used for comparison in the discussion of the archaeology of iron production of Wollega is located to the southwest of the region separated from the first study area by some two hundred kilometers (Burka 2008). As I have already discussed (*cf. chapter II*), lineages inhabiting in both areas claim and count common genealogy, common tradition that their homeland in the northeast close to the Blue Nile. Indigenous iron production practice is one of the areas for which the region is best known. Charcoal making in southwest Wollega made a radical departure from its counterpart in the northeast. Unlike in the northeast, in this region, there is no splendid distinction between charcoal for smelting and smithing.

It is important, however, to note that the forest coverage in both areas is not the same. Unlike in the northeast, here remnants of indigenous plants are only found along streams.

One main tree species used for smelting as well as smithing is *Baddessa*, a preserve for smithing in northeast Wollega. During my ethnoarchaeometallurgical investigation among the Jaarso Abayyani smelters (Burka 2008) of Aira-Gulliso, it was revealed that smiths as well prepare charcoal from eucalyptus trees. This decision might be connected to the high demand for charcoal due to high concentration of smiths as well as lack of hardwood to meet the demands.

In addition, there is no tradition that they could be able to relate to the existence of charcoal preparation in a pit. They knew that their predecessors also prepared charcoal in heaps. Charcoal for smelting and smithing of the reconstructed iron smelting in southwest Wollega was prepared from *Baddessa* (Acacia amythetophylla) tree in an open space. It is possible to offer some explanations, however.

Abee-Dongoro district in northeast Wollega is still relatively under dense natural forest coverage. The Garchi river valley in Abee-Dongoro district has over seventy *gashas* of protected natural forest (Burka 2006; cf. chapter II). Before migration and settlement by groups from the province of Wollo in northern Ethiopia in the late 1960s, (see Waqweyya 1988) the iron smelting lowland part of Abee-Dongoro was densely and widely forested. Slash and burn type of farming has been the major practice (Waqweyya 1988) of the lowland Abee-Dongoro.

On the other hand, Aira-Gulliso district, where widespread iron smelting dominated the economic activity, is virtually deforested (Burka 2008). Iron production might be an explanation for deforestation around Aira town. Few natural forest sites are located mainly along streams intertwined with coffee plants. Probably coffee is the reason for maintaining the forest along the streams. This is because coffee needs canopy trees under which it survives from direct sun light as well as from frost during cold seasons. Coffee forms a significant portion of the economy in Abee-Dongoro lowlands similar to the district of Aira-Gulliso.

Nevertheless, there is a common behavior related to the location where charcoal is prepared in both study areas. In both districts, charcoal is prepared along the streams. The practical and emic explanation seems to arise from the need for water and the labor demanded to fetch for quenching the charcoal. The old smelters express this with such statements as *lagatti galanii; cilattii muratanii* (literally means they settled in the river, cut charcoal). They express this as part of requirements for iron smelting and its location *vis-à-vis* the villages. It is probably because the trees needed for charcoal are located along the streams. It is also possible that charcoal preparation has connections with water for other purposes such as fertility than practical reasons of quenching.

In northeast Wollega, charcoal for re-enactment was prepared near *laga suphee*, literally means Clay River. In the west, charcoal was prepared near a forested stream called *gaawwa*, literally means jungle. Part of the stream system where Baddeessaa trees dominate is known as *laga baddeessaa*, *Baddeessaa River*. It is part of the stream system where smelters and smiths cut wood for charcoal. Today, only the name has survived. Few *Baddessa* trees (Acacia amythetophylla) are left. Inhabitants do not cut the protected trees except with permission from local authorities. We needed to pass through same procedure for the reconstruction purpose.

In the past, there was a song by smelters referring to the decisive role the forest product (*baddessa*) plays by increasing ones financial capacity for asking women for marriage (*cf. chapter V* for a love song by blowers). It tells us a lot about the significance of income from iron production for marriage engagement or man and woman relationship.

Generally, it is the case in Wollega as in other places in the world, that charcoal is prepared either in heaps or in pits (Burka 2006, 2008; Haaland 2004b; Pleiner 2000). The challenge that remains for further discussion is the efficiency of either method of charcoal production. For instance, iron smelting in northeast Wollega took over seven hours whereas in the southwest it took place in about five hours. Local histories of both areas support this in that in northeast Wollega smelting took place once a day while it was twice in Aira-Gulliso area. In addition, whereas bloom is retrieved after it was left to cool down either over the night or during the day time in northeast Wollega, it was withdrawn immediately after blowing was terminated in the southwest.

Nevertheless, further research would probably enlighten us with whether this was due to the nature of charcoal preparation or furnace design or the nature of the ore. For southwest Wollega smelters and smiths, the key factors for lack of diversified use of hardwoods appeared to be necessity as opposed to choices. Lack of choices came from over use of resources and deforestation, which is a characteristic of this particular locality. The smiths' reverting to the use of alien tree species (blue gum, eucalyptus tree) as well as their long time's adaptation to imported scrap iron exemplifies this problem. This shows their flexibility and innovative capability to adapt to new environmentally constrained factors of production.

On the other hand, iron producers of northeast Wollega had various species of hardwood for their own respective choices or for experimenting on each of them. In the face of the scarcity of the woods for charcoal making in the southwest, one could expect the economic use of the resources as well as adaptation to environmental situations. For example, if they prepared charcoal in a pit, it could burn for longer period than that was prepared in heaps on open spaces. It could have reduced the amount of charcoal needed for one smelting. Nevertheless, the open heaps mechanism seems appropriate for closely attending with the making and continuously retrieving of the burned parts of the heap. It would seem that charcoal makers did not want risk their scarce resources.

The pit approach rather took place among the smelters living in a forested environment where varieties of hard wood trees are available. Equally, however, charcoal prepared in the pits has longer durability than the one on the open space. This means that southwestern Wollega smelters could have benefited from charcoal prepared in pit that could redress their shortage of hardwood. As far as I could see, the decisions we could find in both regions show how technological choices are affected by cultural factors. In other words, environmental factors are not sufficient explanations for adopting a certain technology over others.

3.2.2. IRON ORE

Another technological variable constituting the first phase of the operational sequences in traditional iron production is the ore. According to scholars in the field, iron ore exists in various forms (Chirikure 2005; David and Kramer 2001; Hamrla 1964, 1966). Some exist in oxides where metallic iron is bonded to oxygen. Some others found in the form of hydroxides in combination with elements as silica, aluminum, and other impurities. Iron is found associated with igneous, metamorphic, or sedimentary rock formations (Chirikure 2005; Hamrla 1966; Pleiner 2000). African traditional iron smelters exploited different types of iron ores. These include hematite, magnetite, and limonite (Childs and Herbert 2005; Cline 1937; Hamrla 1966; Todd and Charles 1978) or a combination of them. They procured these either in blocks of rock in pits or from sand stones in loose forms (Burka 2006, 2008; Haaland 1985; Haaland *et al* 2004; Hamrla 1966).

In other localities, miners collected the ore from river deposits along the banks in the form of iron fillings (Burka 2006; Sutton 1985). In this last particular instance, they obtain actual ore after washing it out from the other matrices in streams, as in northeast Wollega (Burka 2006). Due to cultural or environmental factors, iron smelters procured and utilized various types of iron ore. One of these is applicable to iron metallurgists in Wollega, its southwest and northeast iron rich regions.

In southwest Wollega, smelters procured iron ores in blocks from outcrops of quartz hill located at Wiinsaa Qachoo, Caagggoo and Warra Qaalluu (Burka 2008; Hamrla 1964/1966). In northeast Wollega, ore was obtained from along deposits of old river channels, by digging narrow and deep pits. They followed the ore veins and then transported the matrices to the nearby streams for cleaning it, as is the case with Fallee mining site (Burka 2006).

According to Hamrla (1966) who conducted field observation in Wollega in 1964 and made microscopic examinations of samples supplemented by data from other areas in the country, the following types of iron ore deposits can be identified in Ethiopia. First, there is metamorphic type, which is of primarily sedimentary in origin but subsequently altered mineralogically and texturally because of regional metamorphism.

The other type of ore deposit is obtained in the form of combined metamorphic, which is a contact metasomatic type because of sedimentation, metamorphism, and metasomatism.[30]This type is formed from decomposition and leaching of extrusive or other ferruginous siliceous rocks. The fourth type Hamrla identifies is hydrothermal type of Katathermal (west Wollega) and telethermal (hot spring) character. Finally, magmatic type is among the constituents. According to Hamrla, however, this type of ore was only known from inference.

Traditional iron mining requires the knowledge of local geology and expertise in the identification of the grains from the matrices. This by itself differed from locality to locality probably based on the availability of the ore at one's disposal. My study of the ethnoarchaeometallurgy of Wollega (southwest and northeast) is very much informative regarding such variability and its various means of exploitation. I have elsewhere, in this dissertation, noted (*cf. chapter II, chapter IV*) that in both study areas of Wollega where

30 It is the process by which the chemical composition of a rock is changed by interaction with fluids or the replacement of one mineral by another without melting. In addition, ores might have been formed in the form of residual concentration.

iron metallurgy has left significant signatures from its past activities, the owners of the skill claim common origin namely close relatives in terms of lineages.

In northeast Wollega (Abee-Dongoro district), there is an ancient mining site today covered with coffee plants and undergrowth at a locality called Faallee, hence *gordana* Faallee. *Gordana*[31] is an Oromo term for iron ore. A granary maker woman, named Balfee Saqoo, married to a Malkee member, is credited for the discovery of the ore in Abee-Dongoro. Smelters who live around the mining sites claim that Balfee Saqoo was the mother of the sub-clan to which they belong. The remnants of pits of unknown depth are evidence for local mining practice since time immemorial. This locality is part of the Garchi river valley formation (*cf. chapter II*). Owing to its nature of formation, the mechanism of procurement must have provided the reason of contrast with its counterparts in southwest Wollega.

In northeast Wollega, iron ore existed in the form of fine-grained fillings (Burka 2006) resulting probably from streams deposition, sedimentation, and pressures. According to old smelters and ore miners, they used to obtain it from an estimated depth of 20 meters following the veins. According to them, the discovery of the ore veins came by a chance. Some pits located near living quarters not far from the major mining sites provided a good example of the confinement of the veins in certain localities and their absence in others. An old experienced custodian of ancestral mining sites accounts that some inhabitants aspired to own the ore close to their houses out of greed and competition.

According to him, the whole Faallee is not generally rich in the ore. There is localized concentration of the ore. This explains reason for the dense concentration of mining pits (Figure 4) in some sites than others. The richest part according to him is where major past exploitation took place. Traces of chains of collapsed trenches and pits evidenced the activity. Today, it is a ritually protected site under canopy trees hanging over coffee plants.

Additional information on the significance of the mines in lowland Dongoro comes from 19th century travelers' accounts on iron items of trade from Horro (Ahmed 1980; Triulzi 1975). Horro is so vast country in northeast Wollega implying the difficulty of exactly locating the source of the item of trade. A local historian by the name of Chernet Waqweyya (1988) made a passing mention of the locality and people responsible for iron production and regional consumption. However, he was the first person to mention that the locality in Abee-Dongoroo called Walagee was the source of iron trade items in the regions of Horro-Guduru. My recent (2009) field works in the highland of Horro also supported the evidence from tradition and

31 As indicated in the text an Oromo term for iron ore is *gordana*. In both separate regions where iron smelting is conducted, iron ore is known by this term. However, there is difference on the terms representing mining. In southwest Wollega (Aira-Gulliso) mining in Oromo language connotes *gordana muruu*, which literally means cutting ore. However, in northeast Wollega (Abee-Dongoro) the Oromo term for mining is *gordana baasuu*, which means to take out the ore. The two different Oromo terminologies for mining among closely related lineages who were engaged with iron production clearly informs the nature of the ore each respective site was dealing with. It is referring to the variability in the identity of the ores in both areas: in the southwest, the ore is obtained in a block of rocks and broken down into pieces before use. In the northeast, the ore is taken out of deep pit with matrices to be cleaned and obtained in filling.

literary sources that the lowland Dongoro has served as a sole supplier of raw iron to smiths in the highlands. On the other hand, informants from Haroo Limmu district close to the Blue Nile in northwestern Wollega recall the presence of ancient smelting over a hundred years. A 90 years old informant, son of one of smiths, recounted the story of local smelting, which he has obtained from his father. That a person of 90 years old did not have personal experience to witness local smelting shows the practice has long disappeared.

However, from the contents of what the informant recounted, it is possible to analyze that the technology of smelting is similar to the one in Abee-Dongoroo. It is difficult, however, to depend fully on the information of few old ages without archaeological investigation to enable wider regional comparison of the distribution of iron smelting technology. In terms of bellows and tuyères, smiths in Haroo-Limmuu carry similar traits with Abee-Dongoro and Aira-Gulliso.

Nevertheless, iron ore mined from such a depth at Faallee comes with unsorted mass of soil. The job of the miners does not stop at the mining pit. When it was retrieved from the pit, there was no way to regard it as proper ore. It was just unsorted mass of soil. It only becomes an ore after they take to the nearby stream and clean it. Cleaning takes place on an acacia bark in the form of panning. The next step is to spread the clean ore on the sun to mildly drying it. If by chance it is too dry, the master smears water over the ore during the smelting. There appears to be a symbolic association between streams, acacia, and ore processing.

There are two things to note about the variables involved in ore processing. First, acacia (*laaftoo*) is a ritual (medicinal) plant in the local community. Its medicinal value has to do with preventing the intervention of malevolent forces such as sorcery during some production activities. Its thorny leaves are also hung in front of the entrance of a house. On the other hand, although acacia is among hard woods for charcoal making, Wollega smelters and smiths do not prefer its charcoal. It is only an important component of firewood in the countryside.

Inhabitants of Abee-Dongoro belonging to Malkee lineage venerate Faallee mining site (Burka 2006, 2008, 2009, 2010). The old[32] miners and smelters who instructed the young smith for the demonstration in Abee-Dongroo belong to the founders of the mining sites. Today, some of them live close to the old mines. According to them, the communities respect the ritual sanction that prohibits cutting trees grown over the old mines. As a result, the old mining site is relatively under canopy trees interspersed with coffee plants. They added that until the mid-1970s when the Derg came with its anti-religious Marxist doctrine, the communities made regularly annual sacrifice of an old cow, *dullacha* to the mining site. Each member of the clan was contributing for the sacrifice. The people also swear in the spirits of the old miners and smelters.

32 One of the Gaagaa brothers who were instrumental for the reenactment of iron metallurgy, Obbo Mosisa Gaaggaa, has recently passed away. The only person left is his brother who has hearing problems. The significance of ethnographic documentation of metallurgy is therefore highly acute. In a non-literate society such as the Oromo of Wollega where there is no tradition to be recorded in written texts, the only reliable source of information is oral history and this is mainly found in few individuals. Few individuals serve as keepers of society's memory.

Another cultural factor of the mining history at Faalle in Abee-Dongoro in northeast Wollega was its association with a woman. The tradition of the clan (Malkee) holds that the founder of iron ore in the locality was a woman who has prepared a pit from which she obtained clay for building granary, *gumbii*. The custom is that women are always responsible for the construction of different sizes of granary immediately before the harvest (Figure 18). In northeast Wollega, there appears no direct meaningful relationship between females and iron production. There is also no physical relationship between granary and iron smelting apparatus. Unlike this, in the southwest, the Oromo term for furnace and granary is the same namely, *gumbii* (see Figure 18) and the furnace is decorated with anthropomorphic features representing a pregnant woman. In contrast, the Oromo term for furnace is similar in meaning with blowing pit, *boollaa buufaa* not with granary. In addition, there is no tradition indicating the practice of storing farm produce in pits.

In southwest Wollega, mining practice involves a separate tradition from its counterpart in northeast Wollega. In the former, iron ore is obtained from outcrops. It required, therefore, an opening of channels and open pits for further exploitation than the exposed outcrops. The ore is transported to smelting site and crushed into pieces. The processes that followed the aftermath of mining also entailed minor actions involving the spread of the ore on the sun near the smelting hut (*cf. chapter V*). Once it is retrieved, no further processing is needed until the day of smelting during which it is pounded on flat stone or timber into smaller pieces. In the past, according to informants, stone hammer was used for pounding the ore. Nevertheless, for the ethnographic reconstruction (2007) they used metal hammer for crushing it on a stone anvil. Stone anvil is also among the possession of the smith at his working place.

Figure 18 Photo and sketch: abandoned granary (left) in a destroyed house of a farmer in northeast Wollega sand its sketch (right). This traditional granary is still in production and use in northeastern Wollega. On the other hand, although there is clear association between the granary and iron production in the southwest of Wollega, the production of such traditional granary is not practiced recently and I did not find it in smelters (or farmers) houses. Only the name existed.

According to old smelters, the ore, for instance, from Wiinsaa Qacoo was obtained through payments for the local landowners, which was more or less similar to the mechanism in northeast Wollega. It is clear from the name of the mining site that in the past a person by the same name appropriated the site. One prominent known owner and bargainer of the ore with miners and smelters was Winsa Qacoo. As a result, the ore-mining site is *gordana* Qacoo. Wiinsaa Qacoo was a noble from the Laaloo lineage. The core of smelters in and around Aira town belongs to Laaloo lineage. Informants recall that the payment was not heavy and was in kind. The owner was said to tolerate people who could not pay and allow them collect without payment. Some also used to enter the site without being noticed, according to informants.

Unlike in northeast Wollega, where the former mining site became ritually important and the lineages continued to protect it as sacred site, the iron-mining site of Wiinsaa Qaacoo is left for pasturing. It follows that the cultural significance of mining sites in southwest Wollega is not salient. It was probably due to the dominance of Protestant Christianity in the region since the turn of the 19th century that has inflicted serious damage on traditional religion. It is a matter of conjecture, however.

3.2.3.FURNACE

Furnace is another technological variable in iron smelting. African traditional iron smelters used a variety of furnaces (Cline 1937; Haaland 1985, 2004; Kense 1985; Schmidt and Avery 1978; Schmidt 1997; Sutton 1985; Tylecote 1975). This is probably to meet the needs of the variety of iron ore in their respective localities or due to cultural reasons. The challenging question is whether such diversity in furnaces came as part of technological response to diversity in iron ore. In spite of this, as I have argued under chapter I, the existence of unprecedented variety in the smelting furnace apparatuses led many scholars to propose the possibility of local experimentation of iron smelting and its multiple origins (Chirikure 2005; Mapunda 2000; Schmidt 1997, 2001, Schmidt and Avery 1978; Schmidt and Childs 1985). How many types of furnaces are there in Africa? What is the criterion to distinguish them into separate parts?

The classification of African furnaces originates from three main features of the apparatus. The first classification looks at the presence or absence of superstructure. It means that whether furnace is limited to underground or built above the ground surface. We must, however, be cautious to observe that no furnace is purely underground without some features above the surface and on the contrary, no furnace is purely above the surface without some features into the ground. However, according to this criterion, there are two types of furnaces. Some African iron smelters used an underground furnace without any superstructure while in other places furnaces with superstructures are used. Thus, bowl or pit furnaces represent furnaces without superstructure. On the other hand, for furnaces with superstructures further category such as shaft or dome or conical shaped ones are made.

Therefore, one of the distinctions of African smelting furnaces is based on the physical features of the apparatus. Such a distinction is used as a parameter to follow the routes of the dissemination of the knowledge (cf. *chapter I* for debates). Combined with models such as migration, such furnace features were used to trace directions of diffusion of the knowledge. This does not explain, however, why people of close cultural identities and affiliated lineages employed two different types of furnaces (e.g. Burka 2006, 2008). For instance, the fact that Oromo smelters in northeast Wollega employed pit furnaces while their close lineages in the southwest used furnaces with superstructure does not conform to the criteria of physical features of furnaces to trace their point of origin or direction of spread (cf. *chapter VII*).

The second criterion for the categorization of African smelting furnaces is concerned with the mechanisms of charging. Smelting is the process of extracting metals from ore matrices in the medium of appropriate heat. Heat is attained in various ways. With regard to iron smelting it is acquired either by forcing air into the furnace or by allowing a chimney means of heat appropriation. Therefore, the question is what types of draughts the various African smelters have employed in their respective localities and to what extent this was useful as evidence for the origin and spread of the technology.

In any case, two forms of draught system serve as criteria to identify for tracing technological similarities or differences (for comparison in Ethiopia, see Figure 19). These include forced draught and natural draught charging mechanisms. Concerning the chimney style furnaces (natural draught furnaces), in some areas secondary small furnaces were installed for further refining of the bloom but in others the unsorted mass of bloom was refined at the smithy. More often than not, the natural draught furnaces are regarded to be large over two meters tall (see Barndon 2004; Cline 1937; Schmidt 1997). There is only one reference to the presence of such type of furnaces in Ethiopia. The Oromo of Shoa in central Ethiopia are associated with an over two meters tall natural draught furnace (Cline 1937).

According to Cline, the Oromo smelters erected the natural draught furnaces at mines and destroyed them after every smelt. Generally, two factors can explain why smelters preferred natural draught furnaces to the other. These are lack of ample labor for charging it or the poor quality of the ore. These are more or less resource-determined factors. It can still be possible that this diversity, but particularly the preference to natural draught furnaces, can be due to either a cultural factor or reflection of groups' style (see Childs 1991).

The other element of the second category is concerned with forced draught furnaces. Forced-draught furnaces are those operated by active human intervention using various types of bellows (see Miller and Van der Merwe 1994; Sutton 1985; Van der Merwe 1980). However, here also further classification is made as in the use of various types of bellows: pot or drum bellows and bag bellows (further discussion below). It generally followed that bag/skin bellows belong to the highest level of the technological achievement in contrast to pot or drum bellows (see Haaland 1985; Todd 1985). The same notion is applied for the comparison between bowel or pit furnaces and furnaces with superstructure. Accordingly, furnaces with superstructures are claimed to be more sophisticated than bowl or similar categories of furnaces such as underground ones.

The third major means of categorizing African furnaces is concerned with the mechanism of separating the bloom from the impurities or *vice versa*. This aspect refers to the complex

Classification attributes						
Locality	By draught system	By structure	By types of bellows	By tuyères direction of penetration	By tapping system	References
Dime	forced	Super structure c. 80cm-1m tall	Pot bellows, with skin as diaphragm	Radiating the base, pairs connecting to the bellows	No design for slag tapping, slag pit, bloom withdrawn from top	Ethnographic observation and record
Oska Dencha	>>	>>	>>	>>	>>	Ethnographic observation and commissioned smelting
Shoa/Oromo	natural	>2m tall superstructure	No bellows	Only two openings, for charging and air-inlet	No tapping mentioned, furnace demolished after each smelt	data retrieval mechanism is not specified
Jimma	forced	About 2m superstructure	Bag bellows from buckskin	Single or pair of tuyères, direction not known	No tapping mechanism mentioned, furnace demolished after each smelt	>>
E. Wollega	>>	About 1.5m 'without' superstructure	Bag bellows from goatskin	From surface downwards, three tuyères in succession	No tapping design, furnace is in slopping position, bloom withdrawn from top	Ethno-history, reenactment of smelting by the writer of this thesis
W. Wollega	>>	>2m tall superstructure	>>	From shoulder back deep into the furnace belly, three tuyères in succession	Slag is not tapped-out, but the mechanism allows the separation of the slag leaving pure bloom at the top	Ethno-history, reconstruction by the help of old smelters, data by the writer of this thesis

Figure 19 iron smelting apparatuses and their classification criteria

activities that take place inside the bloomery furnace. Since the aim of iron smelting is to obtain a metallic iron (see Figure 17), there are different mechanisms for sorting out the impurities. From this, two forms of furnaces are differentiated. The two classifications are known as slag tapping and non-slag tapping furnaces (Haaland 2004a; Kense 1985; Schmidt and Childs 1985; Sutton 1985). This classification is also attributed to the presence of levels in ones technological achievement.

Accordingly, slag-tapping furnaces are placed at the top of achievement with regard to traditional (bloomery) iron smelting. In relation to northeast Africa, its arrival is credited to the Romans (see Shinnie 1985; Tylecote 1975 for analysis of Meroitic smelting technology in the post-1st century AD). With regard to slag-tapping furnaces, smelters have created a special means of extracting the slag out of the furnace so that they could directly obtain a relatively pure metallic bloom from the furnace. Such mechanisms also allowed the air to circulate freely in the furnace since the mechanism avoids slag that could coagulate on the tip-end of the tuyères and gag it. On the other hand, the so-called non-slag tapping furnaces are much complicated and diverse in types. In some areas, furnaces are built on a pit underneath, which is filled with grass and allow the accumulation of the unwanted waste in the pit in the form of slag cake (Haaland 1985; Haaland *et al* 2000, 2004a; Todd and Charles 1978b; Todd 1985). There are other furnaces, which can also be in non-slag tapping category but erected on a flat ground but with slanting position by the help of the slop of the land (Burka 2008) so that the bloom is formed along the tuyères upwards leaving the impurity at the ground surface.

In northeast Africa slag-tapping furnaces are attributed to Roman technology as evidenced by Meroitic smelting technology (see Haaland 1985; Shinnie 1985; Todd 1985; Tylecote 1975). Accordingly, the yet to be researched pre-Roman iron smelting furnace at Meroe appeared to be a bowl one (Tylecote 1975). The question that challenges such criterion of classification is whether it is possible to determine the type of superstructure in the archaeological context. The challenge is particularly acute in sub-Saharan Africa where due to heavy rainfall furnace superstructures are likely to collapse and disintegrate in a short period.

During my second round of ethnographic work in northeast Wollega in 2005 (Burka 2006), I observed that the destruction of technical ceramics under torrential rainfall does not take much time as many smelting sites were built under a shed of tree (see Haaland 2004a; Haaland *et al* 2000) or simple open thatched roof huts (Burka 2006). After the re-enacted smelting of summer 2005 was over, smelters withdrew the clay tuyères and left them outside the open wall hut. During the night of the smelting day, heavy rain fell. In the next morning, when we went back to the site to withdraw the smelting product after it was left to cool down the whole night, we found that the tuyères were literally destroyed. However, this could also be attributed to the level of drying attained by the tuyères makers. The well dried tuyères from past smelting sites could remain under similar weather condition to be recorded after over 50 years of discard (see Burka 2008; cf. chapter IV). However, challenge is very clear for archaeological interpretation of smelting apparatus because of this criterion.

It is archaeologically possible to distinguish between slag pit furnaces and slag tapping furnaces from slag cakes left after each smelt or from tapping channels formed at the base of the furnace (Chirikure 2005). However, there are other furnaces, as I will later on discuss (northeast Wollega smelting), without superstructure and mechanism of slag separation during smelting. It is, therefore, a serious challenge for an archaeologist as well as misleading to claim the possible existence of superstructure after finding only the 'foundation' (see Childs and Herbert 2005) or to differentiate between slag pit and pit-furnace. The furnace that has been used by Walagee smelters of northeast Wollega (Burka 2006) is a different type from classifications into two namely shaft/ dome and bowls. The Walagee furnace is what I called pit-furnace constructed underground in inclined position for over one meter (see Figure 19).

The sizes of furnaces generally depended on the type of technology the group employed, their distinctive culture, and the nature of iron ore obtained in the environs (see Childs and Killcik 1993; David and Kramer 2001; Miller and Very 1987). One of participants in the reconstruction project in southwest Wollega asserts that it is economical from the point of view of maintaining heat to build narrower furnaces than wider ones. He was against the uses of a 50cm radius for the base of the furnace for demonstration.

According to him, less than the suggested diameter was enough. Natural draught furnaces were larger than forced draught furnaces (Sutton 1985). The size explains the amount of labor needed to employ in iron smelting (David and Kramer 2001). It is not possible to operate very large furnaces such as Fipa furnaces in Tanzania using forced draught. The explanation is also related to the quality of the ore that an iron smelter could obtain. It is economical, for example, to use large furnaces for poor quality iron ores to carry out smelting slowly while high grade ores do not need large furnaces to separate the bloom from the impurities (see Chirikure 2005; Schmidt and Childs 1985; Sutton 1985).

3.2.4. TUYÈRES AND BELLOWS

Tuyères and bellows are also significant components of technological variables in traditional iron smelting. The reason why I discuss the two together is to indicate that both bellows and tuyères are often inseparable parts of technological variables in iron smelting. Despite this, for the purpose of detailed discussion, I will take each of them one by one. This also enables to analyze each category with its respective sub-divisions. Thus, I first offer discussions regarding bellows and subsequently take the discussion on tuyères.

There are two main categories of bellows in traditional iron metallurgy- bag bellows and drum (pot) bellows (see Cline 1937; Kense 1985; Todd 1985; also Figure 19). The pot or drum bellows can be made from clay or wooden bowls around which a skin diaphragm is made for propelling air into the furnace (see Haaland 2004a; Todd 1985; Todd and Charles 1978b for the Omotic smelters in southwestern Ethiopia). The bag bellows are mainly made from goatskins although sheep or buckskins are also used (see Cline 1937; Haaland 1985; Huntingford 1955; Lewis 1965; Mcosh 1979). However, there are instances where smelters/smiths have employed the tubes of car tire (Burka 2006, 2008), which signals dynamism in the evolution of bellows.

Nevertheless, there is lack of serious research regarding why smelters employed various bellows. Huntingford's description of materials from which the Oromo smelters of Jimma kingdom made bellows as buckskin seems to lack concern for the cultural significance of the object (see Mcosh 1979). Ethnographic evidence attests that smelters do not simply pick any object for the sake of a fit of its function (Burka 2009). They first had to weigh its purity, the day it was produced as well as the first owner of the object. In case of the bellows, they need to check whether the object was polluted. In addition, the fact that some attributed sheepskin as material for bellows is also challengeable (see Burka 2006, 2008).

According to informants, sheep generally are regarded as 'innocent' and 'dumb' and hence they are not slaughtered as sacrifice during smelting or their skins used for bellows (Burka

2008). The belief is that iron smelting is associated with maneuvers by some 'devil' spirits, whose appeasements offering of goats not sheep. This does not rule out however the significance of necessity for adopting various options for a system to run as the recent use of car tire tubes by smiths/smelters (Burka 2006, 2008; cf. chapter VI).

Tuyères are mainly made from clay though as my current ethnographic work has shown they can also be made other material as iron-sheet and horn (Burka 2006, 2008). Tuyères are as diverse also as the rest of variables in iron smelting. They do not simply vary on materials from which they were made but in the number of tuyères needed and in terms of their sizes. If we take examples from Ethiopian ethnographic records, the Dime iron smelters of the Omo valley of southwestern Ethiopia used 36 tuyères for one smelt (Todd 1985). In the same region, the Tsara smelters of Oska Dencha employed 30 tuyères for a single smelting (Haaland 2004a; Haaland *et al* 2004a). The length of the tuyères in these two smelting sites, however, is not above 30 cm.

In contrast, a report on iron smelters of the Jimma kingdom, in the same river system with the Dime and Oska Dencha but of different linguistic groups, shows that one or two tuyères were used (Huntigford 1955). Further to the northwest of Jimma, in Wollega the number of tuyères needed for one smelt is ten. Apart from the permanently installed horn and iron tuyères, which are about 10cm in length outside the nozzle of the bellows, clay tuyères of Wollega are generally long. However, it is difficult to standardize the length of the clay tuyères as the determining factor lies in the size of each furnace and the direction and depth of penetration into the furnace. Accordingly, when the furnace is about 80 cm in depth (as in northeastern Wollega pit furnace for re-enactment), the length of the clay tuyere is about 70cm to leave a difference of about 10cm between its tip ends and the ground floor of the furnace (Burka 2006). However, with regard to the tall furnace of southwestern Wollega, the length of the tuyère increases accordingly.

Accordingly, for the ethnographic reconstruction of the smelt in one of the two sites in southwest Wollega, the clay tuyères were about 150cm long meant to fit to the 160cm tall furnace (Burka 2008). The measurement is between the opening through which the tuyères were inserted and the floor of the furnace over which they are hanging, close to the rake channel. The depth of penetration into the belly of the furnace both by Wollega smelters in southwest and northeast is the same. The fact that clay tuyères were inserted deep inside the furnace shows how the society conceptualizes effective procreation. In other words, the deeper the penetration, the higher the probability of fertility is.

The other variability in tuyères is the direction of insertion into the furnace. In many instances, tuyères are inserted (for both natural and forced draught) at the base of the furnace (Barndon 2004; Haaland 2004a; Haaland *et al* 2000; Schmidt 1997; Todd 1985). In a few instances such as in Cameroon, however, tuyères are inserted from above the mid-section of the furnaces (Burka 2008; Childs and Killick 1993; Killick 1999, 2009). The direction of insertion and the depth inside the furnace is also another form of technological variable but also needs to be examined in accordance with the cultural milieu in which it is operative.

Nevertheless, the most arguable aspect is regarding the technological significance attributed to the depth of penetration by tuyères in African bloomery furnaces (Killick

1999, 2004; Mapunda 2000; Schmidt 1997; Schmidt and Avery 1978; Schmidt and Childs 1985; Schmidt and Mapunda 1997; Todd and Charles 1978b). This argument has not yet been settled. According to its proponents (e.g. Schmidt 1997; Schmidt and Avery 1978; see also Todd1985), the fact that African smelters inserted their tuyères deep into the furnace is said to have allowed what is known as 'preheating' of air before it enters the combustion zone inside the furnace.

Based on this argument, for example, Todd (1985) attempted to establish similarity between the Haya smelting technologies in Tanzania and her study area of the Dime of Ethiopia. According to Judith Todd (see also Todd and Charles 1978b), if for example, smelting normally takes place at about 1400°C, then due to the long tuyères inserted deep into the furnace that allows 'preheating' of the air in the long tuyères, direct reduction begins to take place at about 800°C. In other words, through the mechanism of inserting tuyères inside the furnace, African smelters might have obtained an advantage of about 600°C. This resolves the problem of attaining high temperature as proponents of outside origin argue (see Phillipson 1985, 2005 on problems regarding temperature).

One of the research questions I have posed (cf. *Chapter I*) in the search for technological significance of the Wollega tuyères was whether the size of the tuyère and its deep insertion enlightens us on the debatable 'preheating' technology. As I have noted (cf. *chapter V*), the Wollega clay tuyères are fully installed inside the furnaces of the southwest and northeast (Burka 2006, 2008). In addition, I have observed that the flame of the fire was bouncing back through the tuyères openings indicating that the air inside the tuyère is generally hot (Burka 2008). This appears to pose a number of challenges. The problem of flame that comes back through the clay tuyères forces blowers to continuously pumping without interruption (cf. *Chapter V*). The problem becomes very clear during the time when the blowers were replacing one another.

The philosophy behind 'preheating' is that the air inside the tuyères was heated before it lands on the charcoal and ore inside the furnace. Since air is one of the variables in iron smelting, its manipulation by the master smelter including through the design of the tuyère or its level of penetration inside the furnace remains to be further investigated. In other words, it needs rigorous experimental observation before endorsing or rejecting the assumption (e.g. Killick 1999). According to some scholars the theory of 'preheating' technology said to have been developed by African smelters (see Schmidt and Avery 1978; Schmidt and Childs 1985; Schmidt and Mapunda 1997) long before their European counterparts should await serious laboratory examination through experimental smelting (Killick 1999).

Generally, however, various smelters among different African societies used diverse sizes of tuyères. In some cases, they have also used different types of tuyères (cf. *chapter V*). African smelters also varied in the number of tuyères used for single smelt. They have also differed on the direction of penetration of the tuyères into the furnace. The question that remains is whether this diversity can be attributed to technological factor or cultural preference of each society. It is clear from the names given to the various tuyères that diversity due to respective cultures of each society is inevitable. In the following (chapters IV &V), I will concentrate on the description of survey of archaeometallurgical sites (chapter IV) and ethnoarchaeometallurgical case studies from the two study areas in Wollega.

CHAPTER FOUR

4.0. THE ARCHAEOMETALLURGICAL RECORD: SURVEY AND RECORD OF VESTIGES OF IRON PRODUCTION IN WOLLEGA

4.1. INTRODUCTION

The archaeology of iron production is the study and analysis of vestiges of metallurgical activities. Vestiges of iron metallurgy include a number of components required for the production of the desired material. Vestiges are left according to the *chaîne opératoire* of production trajectory. These, for instance, include ore mining and processing sites, iron-smelting sites with possible remains of furnace fragments, of slag, tuyères, charcoal, ore fragments, clay processing objects, and unprocessed bloom fragments. As regards smithing sites (e.g. metal pieces), structures (e.g. hearth, house foundation, and pits), stone /metal anvils upon which either bloom was refined etc are generally the components of archaeometallurgical investigation. Stone anvils can either be located close to the smelting site or within the compound of the smiths. However, this depends on the nature of the structure of the society under study namely whether iron smelting and smithing is under the same occupational identity or not.

Based on the level of social organization (stratification and specialization) of a given society, types of iron production would take these various forms and understood accordingly. In some cases, not only are the smelting and smithing sites located at a very close range (within the same village), but also the actors of both practices would be found in single personality or a family (see for instance, Haaland 1985, 2004a; Haaland *et al* 2000, 2004b; Todd 1985; Todd and Charles 1978a, 1978b in Ethiopian Omo valley). This would complicate the attempt of investigation and demands auxiliary scientific tools and ethno-history of the region for identification of one phase of production from another in the *chaîne opératoire* of iron metallurgy. The complication is so strong in that remains from smelting can overlap with that of smithing or roasted ore can be confused with half cooked ones (see Alpern 2005; Chirikure 2005; Killick 1999, 2004; Pleiner 1980. 2000; Tylecote 1975, 1980).

In others, the two activities not only belong to two different specialized communities, but they are situated in separate geographical settings (e.g. Burka 2006, 2008, 2009, 2010 in Wollega; Huntingford 1955 and Lewis 1965 in the kingdom of Jimma Abba Jifar). A related question must be the archaeological significance of such settings for the interpretation of the remains. However, such a distinction should also take into account the fact that specialization does not necessarily entail differences in lineage relationships. The distance between these two occupational entities could vary from a few kilometers to tens of them depending on the spread of artisans across spaces particularly where resources concentrate. Therefore, the level of social and political complexity attained by the group under study certainly affects the nature of survey result of archaeometallurgical vestiges across traditional iron production sites (see Haaland and Haaland 2007).

The Wollega archaeometallurgical evidence is in the category of the latter. The discussion in this chapter, therefore, gears towards distinguishing the challenges and opportunities associated with the investigation of the archaeometallurgical record as practiced by different artisan groups (smelters and smiths). As we have noted in different sections of the dissertation (cf. chapter II), traditional iron production in Wollega constituted two different socio-economic categories namely smelters and smiths. These two occupational categories were clearly distinguished from each other-both in terms of socio-economic, and work and habitation sites. These distinctive features include the fact that smelters and smiths belong to separate social and economic divisions including lineage or sub-lineage variations (*cf. chapter II*).

In this context, the two occupational identities (*cf. chapter VI*) occupied or settled in separate ecological niches. This is well established from ethnographic, ethno-historical, and archaeometallurgical evidence (see Burka 2006, 2008). Accordingly, iron smelters and smiths in the districts of northeast Wollega (Abee-Dongoro) and southwest Wollega (Aira-Gulliso) not only belonged to separate identities (occupational and social) but also inhabited different and separate villages/hamlets of their own (Burka 2006, 2008). These habitation sites are distinguishable in the archaeometallurgical landscapes as creeks or river streams or some sizeable forest often separates them from one another (*cf. Chapter II*). This is particularly clearer in northeast Wollega, Abee-Dongoro district, where many streams dissect hamlets in the study area.

Consequently, it is plausible to posit that smelting and smithing sites could easily be distinguishable from one another in the landscape (see Burka 2006). My argument in this regard is that, systematic survey, recording of archaeometallurgical signatures in former production sites, would significantly contribute to additional information to enhance our scientific knowledge about the activities, and the behavior of their actors embedded in the remains. This would probably be enlightened when we are able to support it with ethnoarchaeometallurgical evidence (cf. chapter V).

The challenge that remains, however, is whether survey of archaeometallurgical vestiges could provide ample evidence on how the activities took place i.e. the whole processes that should constitute iron metallurgy. As a result, combination of various methods of retrieving data (ethnographical, ethno-historical, surface observation and collection, shovel test of selected site) was employed (*cf. Chapters II, III, V*) in order to deepen and widen our perspectives. It is hoped to fill the gaps left out due to lack of access to human decisions during the activities in the past. The other important contribution to our knowledge is gained because, in Wollega, some former smelters/smiths or participants of traditional iron metallurgy or descendants of former smelters/smiths are still available to provide relatively first hand information on the nature of past activities.

In other words, the fact that traditional iron smelting practices have continued to the 1960s makes our endeavor, for accessing the information contained in remains, extremely exceptional and meaningful. This, however, is not to overlook the fact that change might have occurred throughout time and across space. On the other hand, these opportunities were among the impetuses that enabled me distinguish between smelters and smiths as well as smelting sites *vis-à-vis* smithing sites including the rest of variables in iron production.

Some of the limitations in achieving the goals of this chapter and some of the objectives I have set out for include, however, the lack of archaeometallurgical tools (mainly laboratory and skill) for analysis of surface finds. Otherwise, the analysis of e.g. slag fragments, slagged tuyère tips, furnace debris could have helped distinguishing whether the archaeometallurgical remains belonged to past smelting or smithing activities.

Thus, survey of the components of traditional iron metallurgy provides mainly spatial information on past activity areas as well as their temporal aspects when supplemented with other aforementioned data retrieving techniques. For instance, iron ore mining sites in southwest Wollega is widespread across the districts of Yubdo, Aira, and Gulliso at localities including Dimma, Wiinsaa Kacho, Warra Kallu, and Chago mineshaft sites. Smelters were said to have explored and exploited the aforementioned ore quarry areas over a period throughout the region (cf. chapter II).

Although it is specifically difficult to confidently outlining dates for the various ore quarry areas, it is possible to forward an assumption. Yubdo (Dimma) iron ore mining site might have been the oldest of all. One main reason for assuming the argument is its association with traditional gold prospection activities since indeterminate period in the past (see Cerulli 1933). This is an area leased to prospectors since the turn of the 20th century. In addition to this, according to local history, the Italians attempted to further developing its infrastructure during occupation period in the late 1930s. It also partly explains why local mining and smelting activity began to dwindle during this particular period. Firstly, because Yubdo was a project area for gold prospection at grand level, therefore local miners were discouraged. Secondly, by supplying with remains of car leaves to sustain local demand for metals, the Italians also diverted labor to gold/platinum mining and processing than iron metallurgy.

From this, it is likely to posit that local iron ore prospection might have started and developed as part of gold prospection and panning practices. In other words, there possibly was pyrotechnological background as they might have experimented with gold metallurgy. Then, miners seemed to have gradually identified outcrops in various parts of the districts, first around Winsa Kacho and then northwest wards to Warra Kallu and Chago ore outcrops in the district of Gulliso. The other possible scenario was that prospection of iron ore could have simultaneously developed all over the aforementioned sites. However, the discovery of mining sites at Kacho and Chago could be associated with the change of the status of the Yubdo district mining site from local level to attract national and international interests since the turn of the 20th century when the Imperial government leased Yubdo area to platinum prospectors, which the locals called black gold (see Hamrla 1966).

The discovery of Winsa Kacho mining site in particular might have been a turning point in the regional traditional metallurgy. This is because the major concentration of iron smelters and smiths is found around Aira town. The smelting and smithing core area is about 10 km east of the town close to Birbir River along which are also located Dimma, Warra Qaalluu and Chago/Caaggoo ore mining sites. On the other hand, ethno-historical sources also confirm about the significance of mining scars left across the surfaces of the aforementioned localities. Iron (in the form of raw material or finished products) from these various localities might have circulated across wider areas in Wollega and beyond (Ahmad 1999; Triulzi 1975) since antiquity as recorded by travelers in the 19th century (see Kenno 2007).

On the contrary, as far as current record is concerned, iron smelters in Abee-Dongoro district in northeastern Wollega seemed only to have depended on a single iron ore-mining site. Only Faallee mining site is known to have been exploited by local smelters and traded outside of Abee-Dongoro (Burka 2006). Oral histories attest that the ore has first been tried outside Abee-Dongoro and the smelting seems to have been started after the news of its rich content was attested. The other peculiarity of Faallee iron ore was its ownership and the processing mechanism. Unlike in southwest Wollega, where smelters would enter into contracts with landowners away from their production areas, the smelters of Abee-Dongoro were owners of ore-mining sites. It explains the regional confinement of iron smelting in Abee-Dongoro to the lowland parts of the district.

In other words, smelters of Abee-Dongoro have once controlled the flow of iron ore, raw iron (bloom) and finished products (in this case, since it seems they have also controlled the smiths). In this case, the distribution of smelting sites was across the spaces inhabited by the groups who controlled the mining sites (*cf. chapter II*). According to ethnographic evidence, former mining site is located relatively close to habitation quarters occupied by descendants of former smelters and miners. In other words, hamlets where descendants of iron smelters and owners of the mining site live are situated between the hamlets of smiths and the mining site.

Ethnographic and ethno-historical evidence in northeast Wollega, further on, ascertains that within close families, there were people who were experts in mining while the rest were smelters (Burka 2006). Other classes of the same lineage were in charge of the distribution of the wealth and reciprocated with various forms of services such as working on bellows. Faallee miners used to refine the ore from the soil matrixes at an immediate river located between the mining site and the hamlets of the smelters (the clan chiefs) and of the smiths in a mechanism similar to panning gold. However, there is no gold prospection practice in Abee-Dongoroo in general. Interestingly enough, however, former smelters used to call iron ore 'gold of Faallee'.

Survey of the archaeological record of traditional iron production practices in Wollega has also increased our knowledge regarding the settlement pattern connected with various activities. These include sites associated with mining, settlements of iron smelters, villages where smiths concentrate and the mainstream community in general. Accordingly, archaeological survey of these sites in both southwest and northeast Wollega has differentiated between these sites. The settlement pattern is much clearer in northeast Wollega due probably to the natural boundaries created by creeks. From the surveys, it became clear that iron ore mining sites are generally located outside areas inhabited by smelters and smiths as well as the rest of the community.

The exception to this general rule is the Faallee iron-mining site in northeast Wollega. As previously indicated (*cf. chapter II*) practices in the Faallee mining site were peculiar (Burka 2006) because some family members of the smelters claimed to have ownership right and lived close to the sites. The location of habitation sites by members of Malkee lineage probably rooted in the need for control of the resource. Accordingly, due partly to the forested streams that physically separated hamlets and partly to the fact that the chiefly members of the lineage, to whom ultimately ownership over resources belongs, lived at a fair distance from the mining sites around which smelters concentrated, it is possible to evidently distinguish between sites belonging to smelting from smithing sites. Such knowledge of local histories enabled the identification and documentation of archaeometallurgical sites in both regions of Wollega.

In the following section, I will outline sites associated with former iron production in southwest and northeast Wollega. In doing so, I will concentrate, primarily on the description of former smelting sites. As far as the nature of iron production is concerned, smelting is central to traditional metallurgy. Since smelting is at the core of iron metallurgy and this study, it is believed to address the rest of components of traditional iron production from this vantage point. In spite of this, I will try to balance the emphasis of areas of discussion by recording parts of the *chaîne opératoire* of iron metallurgy.

4.2. THE ACHAEOMETALLURGICAL RECORD

4.2.1. SURVEY OF MINING SITES

At the beginning of the *chaîne opératoire* of iron production is iron ore, the major ingredient of metallurgy (*cf. Chapter III*). As I have discussed previously, there were various ways of obtaining the ore (*cf. Chapter II*). This depends on various factors. The reason for diversity in prospection can be found either in the nature of local geology or culture of the local society or the technological knowledge of mining (see Burka 2006, 2008). This is evident from survey record of former traditional iron ore mining practices in northeast and southwest Wollega. Two different data areavailable for mining practices in Wollega. The categories are rather geographical but also dependent on local geology of the respective zones.

Thus, we can categorize Wollega mining styles into northeast and southwest ones. Why were the styles different? Was it due to cultural matters? In other words, were the two miners unknown to each other? These and other related questions are discussed in different parts of the dissertation. The objective of this chapter is mainly to record, classify, and analyze the sites than to offer interpretations. There seems no doubt, however, that in every description there is a certain amount of interpretation.

Variability in terms of mining style in the two Wollega regions (northeast and southwest) seems to have been determined by the nature of the ore. In northeast Wollega, Faallee mining site (Figure 20), the ore was obtained mainly in filing (powdered) form. It was obtained after the soil matrix iscleaned at a stream in similar fashion as traditional gold panning practice. On the contrary, survey and examination of mining sites in southwest Wollega revealed that the ore was in its block-rock form, which was knocked-off at the site and crushed into smaller pieces at the smelting site.

Accordingly, in northeast Wollega, the ore is obtained from a narrow pit (Figure 20). The pit is a vertical shaft dug deep into the ground. The depth of each pit is estimated, according to former miners, to about twenty meters. However, due to a number of natural and cultural factors it was not possible to examine the depth of the pits and prove or disprove the claims

Figure 20 one of narrow mining pits at Faallee, Abee-Dongoroo, estimated to about 20m in depth

(for details cf. *chapter VII*). One of the barriers was the fact that many pits have collapsed and formed long channels while those that look intact could not be trusted.

As a result, I could not risk people to examine its inside. Even then, there was no person who volunteered to enter. In fact, in addition to its physical fragility (Figure 20) fear of possible infestation with wild creatures such as pythons or snakes over the years since miners abandoned the practice. Other cultural matter was the fact that the present status of the site is different. It became ritualized and highly revered. They seemed to have regarded it as though they would face possible supernatural sanctions if they disturbed the sites. The only thing they did was to show me around, and take the ore for demonstration from surface close to one of the pits and depart.

Survey and surface examination of former mining sites at Faallee locality (Burka 2006) has provided significant archaeological evidence, which reveal intensive human actions in the area in the past (Figures 20, 21). Typical of this site was that narrow pits only allowing a single person to enter at once are distributed under the bushes and undergrowths (Figure 21), which have reduced visibility and surface recording. Too many of the shafts were fairly distributed across the currently forested and revered area but not too close to one another.

About one hundred pits were located a few meters from one another. Many of these shafts have already collapsed and formed long channels. Without information obtained from old

Figure 21 One of old mining shafts, half collapsed and covered with undergrowth

miners and observation of intact pits, an outsider can wrongly conclude that channels were primarily opened for mining purpose. During the fieldwork, one of the two survivors of former miners, those who claim descent from the senior *balbalaa* (cf. *chapters II, V*) of the Malkee lineage, frequently cautioned the team from stepping on one of the collapsed pits covered with undergrowth. It is too dangerous to walk through old mining sites without a guide from local residents particularly the 'keepers' of the ancestral site.

The fact that pits have collapsed from the surface was probably that they were connected underground and that they were not as deep as former miners have claimed. Today, canopies and coffee plants covered former mining pits. According to two surviving former ore miners, one of them could not communicate with us due partly to hearing problem, once the depth where the ore vein begins was attained, excavation was conducted following the 'veins', often horizontally.

Consequently, many independent pits have become connected under the ground and were cause of disputes among miners. It was probably due to this reason that pits have collapsed and formed a channel, which is covered with undergrowth (Figure 21). As a result, the current observation of long channels in the mining field should not tempt us to conclude that former smelters had created it. In other words, the present status of mining sites could affect the nature of archaeological interpretation of the mechanisms adopted by ancient miners.

According to ethnographic record, former mining sites at Faallee are restricted to human and animal access. The mining pits frequently entrapped and endangered domestic animals. Nevertheless, restriction on human access, including prohibition of cutting trees, has a rather different reason than physical fragility of the pits due to old age. According to former miners, smelters and owners of the land, the sites have become ancestral ritual sites revered by members of the Malkee lineage (Burka 2006, 2009). The belief is that the spirits of former miners and founders hover over and reside in ancestral mining sites under the forest.

Formal practice of worship and annual offering at the site is said to have continued up to the 1970s and discouraged afterwards. Nevertheless, during our fieldwork in 2005, I could record from different informants that the site was still revered and protected from intervention. One of the ancestral sprits called for facilitating smelting was that of Balfee Saqoo. Balfee Saqoo is the legendary woman attributed to the discovery of the ore at Faallee during her routine granary making. The difficulty is however to discern, from archaeological point of view, ritual significance of the site. Intensive examination of the mining sites in the future, if allowed by the keepers of the ancestral site, might yield evidence of ideological significance of the site from discards of the offerings (cf. chapter V). However, since workers were possibly served with food during the active mining activities, the distinction between remains of the secular from ritual would be difficult.

Figure 22 one of over 86 circular pits recorded at Wiinsaa Qacoo, east of Aira town. It is the closest mining site for smelters around the present town of Aira, of high concentration of former smelting sites.

On the other hand, as I have previously indicated in this chapter and elsewhere in this dissertation, iron production practices in southwest Wollega depart distinctly from similar activities in northeast part of the region. Accordingly, one can begin to distinguish the differences from the mining practices. To begin with, southwest Wollega miners had various options for ore prospecting. Thus, they were not confined to one particular site and were not determined by factors such as scarcity of ore resources.

Despite similarity in local geology and that, the quality of the ore was more or less the same (*cf. chapter II*, Hamrla 1964/1966); miners have explored various sites over a period. From a survey of three former mining sites (Wiinsaa Qacoo, Warra Qaalluu and Caaggoo), it became clear that people of similar mining technological expertise might have been involved in its activity across the region and throughout time (see Figures 22, 23 and 24). This could probably have been determined by the nature of local geology. The ore of southwestern Wollega was obtained from exposed outcrops in the form of blocks of rocks than iron filings in the northeast Fallee mining site of Abee-Dongoroo.

On the other hand, the nature of opening up of the pits is generally different (see Figure 22). As opposed to the one in northeast Wollega where mining pits are not only deep but also dug narrowly to allow only one person to enter (Figure 20), in southwest Wollega (e.g.

Figure 23 A mining pit, which is half field with foliage, is one of many scars in the bushes at Warra Qaallu traditional mining site, located at about five kilo meters southeast of Gulliso town

Figure 24 this is a mining pit at Chago. Gordana Caaggoo is highly reputed and as widely exploited as the site at Wiinsaa Qacoo Note the exposed outcrops of ore on the surface close to abandoned pit.

Wiinsaa Kacho) the mining pits were widely opened (Figure 22). The examination of the size of the pit at Wiinsaa Kacho shows that ore at the site was close to the surface because of which miners probably focused in lateral excavation than going deeper vertically. The average depth of one of the larger pits is eight meters. However, this should take into consideration the amount of change on the site in the last over forty or fifty years.

From the location of the mining sites *vis-à-vis* the major concentration of smelters inside and around the present town, one can wrongly conclude that iron ore has only been obtained from Wiinsaa Qacoo. However, oral and ethno-histories attest that smelters had opportunities and choices of obtaining ores from various localities including from Yubdo district at Dimma. It seems, however, that smelters close to the border of Yubdo district were much inclined to ore prospection from Yubdo area while smelters of Aira town area have memories of association with Wiinsaa Qachoo than Yubdo.

Similarly, former smelting sites were recorded close to another mining site at Warra Qaalluu (see Hamrla 1966) during September/October 2010 field survey (see Figure 23). Warra Qaalluu mining site is located on the edge of a stream, a tributary of Birbir River, where locals continued to pan gold by traditional means. Despite its close location to a gold prospection area, unlike the northeastern Wollega ore processing practice that looked like gold panning one, at Warra Qaalluu ore was obtained in blocks of rock. It is similar to practices in other nearby sites in southwest Wollega namely at Dimma, Wiinsaa Qacoo, and Caaggoo.

It is important to note that the mechanism of mining in both areas in Aira-Gulliso seems to be similar as far as the mining scar shows. In both areas, separated by a few kilometers of distance, miners have used a circular means of digging out the outcrops. The third widely exploited mining site around Aira-Gulliso is known by the name of Chago (Hamrla 1964/1966).

The local pronunciation of the site was a reason for me to fail to locate and conduct survey of the site it until September/October 2010. Reports at the archives of Ministry of Mines and Energy mention about the presence of former local mining sites (see Harmla 1964/1966). Locally (unlike its sound in English), it is pronounced as Caaggoo with explosive C and long sound. It was because of this that local people could not locate a mining site by the name of Chago (see Figure 24).

Iron ore at Caaggoo is visible on the surface of the ground than elsewhere in other sites. Probably partly, due to land degradation that was caused by intensive mining practices, iron ore is enormously scattered over the surfaces and the outcrop is exposed on to the surface (see Figure 24). From the circular pits across the landscape, it is possible to conclude that the technological knowledge, which miners adopted, was entirely similar to the rest of the sites in Aira-Gulliso district. The traditional mining practices were markedly different from modern techniques as evidenced from the long channels created by teams from the Ministry of Mines during exploration in the late 1950s and early 1960s (e.g. Hamrla 1964/1966) to determine its economic feasibility. Without ethno-historical evidence, the observation of the channels in the mining landscape could complicate archaeometallurgical interpretation on the nature of mining techniques adopted by the people.

In addition, a widespread signature of human activities on the landscape at Chago (Caaggoo), some five kilometers southeast of Gulliso town (see Hamrla 1966) shows that the region has been one of the highly exploited areas before indigenous smelting was dropped in the second half of the twentieth century. It might have also served the need of smelters in the immediate environs as evidenced from vestiges of iron production activities.

4.2.2. A SURVEY OF PAST SMELTING SITES

As I have already mentioned, traditional iron production practices offer a glimpse of how people in the past interacted with their environment and among themselves. In the *chaîne opératoire* of iron production practices, smelting is a critical stage. At the smelting phase, various inputs such as the bloom and the slag (*cf. Chapter III*) as well as the furnace and other refractory ceramics are put together for the realization of the change from a mostly natural aspect to one of cultural products. Partly due to this and partly due to the fact that smelting sites are relatively permanent family sites, rich information on the nature of human decision and behavior is readily available for site investigation and offering plausible interpretation.

The location of smelting sites *vis-à-vis* the human settlement sites and of mining areas offers insights into the nature of interactions among humans as well as human interactions with their environment. The relative distance of smelting sites from habitation sites can offer insight into the nature of space utilization by people in the past. However, the problem is whether ethnographic evidence could inform us this since former smelting sites are today located inside the residential quarters. One of the objectives of this study as indicated (cf. *chapter I*) was to distinguish between various specializations sites in traditional iron production as well as whether sites were located within the villages or on the margin of them. As we have noted before in Wollega, smelting was generally a separate occupational specialization carried out in separate locality away from smithing sites.

As regards the smelters, they also occupied a separate occupational and social identity from the smiths. They claim to belong to the mainstream society. I have also indicated the simplicity of this statement in the face of socio-cultural dynamism through which the societies pass (Burka 2009). We have evidence of separate incidents in the post 1950s. Since the 1940s, traditional smelting was diminishing in both northeast and southwest Wollega. According to informants from both regions, the information of which is generally similar, former smiths have crossed the boundary of occupational identity to practice smelting. What is the archaeometallurgical significance of this story?

This implies that some smelting sites could be located close to the hamlets where only smiths and remains of their activities should be. Similarly, it was discovered that a family of smelters from the *Malkee* lineage in southwest Wollega have crossed the boundary by settling inside the villages belonging to *Innawangii*, *Innanani*, *Abayyanii* lineages specializing in smithing. Some individuals from *Innawangii* and *Innabalxii* group of smiths were trained how to smelt and could build their own smelting houses within the territory of the hamlets of smiths (Burka 2008).

A similar incident happened in northeast Wollega (Burka 2006) where members of Yamalagii lineage, most of whom were smiths, conducted mining and smelting activities. According to informants, this happened during the last phase of traditional smelting practices. Nevertheless, one could not be assured whether crossing boundaries of occupational identities was confined to the period when legitimate smelters decided to quit the activity. The family of Kumsa Libdi (chief of the smiths) was connected with this incidence in northeast Wollega.

According to the local history, Kumsa Libdi used to conduct mining at Faallee. Former miner from Malkee lineage associates this with the period with dwindling in the regular smelting practice in the post-Second World War period. It is interesting to note, however, that Libdii has beenpromoted to a chiefly class in the second half of the 19th century when northeast Wollega fell under the Amhara rulers from Gojjam (cf. chapter II).

Conversely, in the end, after smelting has been dropped in the southwest, some members of the Malkee lineage took up smithing from their former clients and apprentices and today their descendants are occupational smiths. Today, for an outside observer, the family is just one of the smithing families of the village known as Abayyanii of about 54 households. In reality, however, they can only be distinguished as members of the smiths who occupied

Figure 25 one of the family members of the owners of smelting site in Walagee locality of Abee-Dongoroo was standing over former family smelting site. Note that the place is now forested and visibility of the mound is not possible due to the undergrowth.

the profession through learning.[33] In terms of social identities and therefore commensality problems, they claim that they still enjoy equal status with members of the mainstream societies. The boundary between members of the mainstream and the smelters is generally illusive. This is unlike those individuals or groups who belong to the social identity of the smiths.

Contrary to the claim by smelters, my observation of the livelihood of the wife of a former smelter who settled among the smiths shows that they were potters (Figure 26). However, it is possible that the social status of the former smelter was changed due to the new position he took while living inside the smiths. The other possibility was that the former smelter was dictated by lack of wealth to reestablish his position in the previous lineage by marrying family of respected status. On the other hand, this was probably because the

33 I have an uncle who learned how to weave clothes when he was in prison during the Imperial period. After he was released from prison, he took up weaving as part of his livelihood. Although many relatives were embarrassed by his new job and tried to ridicule him, he however maintained equal status with the rest of members of his family. In other words, he has never been treated on equal basis with the *tumtuu* artisans, the socially weavers.

Figure 26 this is a clay soil at the stock of the potter wife of the son of former smelter. Interview with her mother-in-law also revealed that she was a potter too before she stopped due to old age.

smelter himself from the Malkee lineage might originally been low class. In other words, he might have been, for instance, a smith who learned smelting from his masters and moved to the hamlets of smiths as an itinerant.

On the other hand, survey of past smelting sites in northeast and southwest Wollega has provided some challenges and opportunities. The opportunity was to be able to distinguish between hamlets connected with smelting and smithing. The challenge has to do with the type of environmental context in which the activity areas were situated. I have previously discussed (*cf. Chapter II*) that northwestern Wollega offered unrivaled opportunities and choices for charcoal makers due to its rich forest coverage in the district of Abee-Dongoroo and its environs extending to the highlands of Horro. This, with undergrowth around abandoned smelting areas (e.g. Figure 25), impinged on survey and record of vestiges of iron smelting.

At a distance of about 20 minutes walk from this smelting site in Abee-Dongoroo (Figure 25), we were able to locate another site for smelting (Figure 27) with the help of members of the family ofthe former owners. This too was problematic due to undergrowth, which is waist deep and thick. We were only able to identify some remains of smelting activities such as slag fragments (Burka 2006).

Figure 27 Note that remains of smelting activities were made possible by pushing the undergrowth from the site identified by another son of former owners.

The positive input of ethnoarchaeology of iron smelting in Wollega, as I have discussed elsewhere, is the presence of survivors of members of the smelters community, which was important on several grounds. First, they served as guides to locate former activity areas and their owners. They also helped in differentiating between various types of specialty in iron production-smelting and smithing-as well as other activities including furnace building skills. The other significance is related to iron production sites in southwest Wollega. Firstly, in contrast to northeast Wollega, iron smelting has been widespread in various localities in the districts of Aira and Gulliso. Secondly, due to the sparse vegetation cover and highly exploited environment, it was much easier to locate the sites where its vestiges were partly exposed and visible. The other important input with regard to southwest Wollega iron metallurgy sites was its rich memories among the inhabitants of the region.

In addition, one set of former smelting localities in the region was located in the present day Aira town, which originated as a missionary station. The recent developments encroaching on former smelting sites make the record of iron smelting activities in southwest Wollega more important. Furthermore, of all the smelting sites surveyed, the present day Aira town locality was essential in the sense that it was on the one hand a highly concentrated metallurgical area and on the other hand, many former smelting remnants are now exposed due to construction activities. Some of these former smelting sites are located in the compounds of individuals who did not know anything about the history of former production activities (Figure 28).

Obbo Oliqa Lamu, son of former smelter inside the present town describes the occupant (Figure 28) as newcomer who does not have any link with the tradition. According to the owner of the compound where the vestiges are exposed, it was always intriguing her that the compound was generally immune from termite infestation. Future research would probably enlighten us whether archaeometallurgical remains could serve as medicine to get ride with termites.

Aira town locality is one of the richest former smelting sites, which I have visited, photographed and took the coordinates (*see map of Iron Age sites in Aira Gulliso, Chapter II*). The reasons for identification of the site are generally two. The first is that oral histories and the nature of the remains are useful to distinguish the type of activity in each site. The other importance is that it enables to identify the owners of the sites. Only one smithing house is recorded on the edge of the town.

Overall, Aira town and its closest villages are only associated with smelting but not smithing. On the other hand, there is a clear difference between sites associated with smelting and

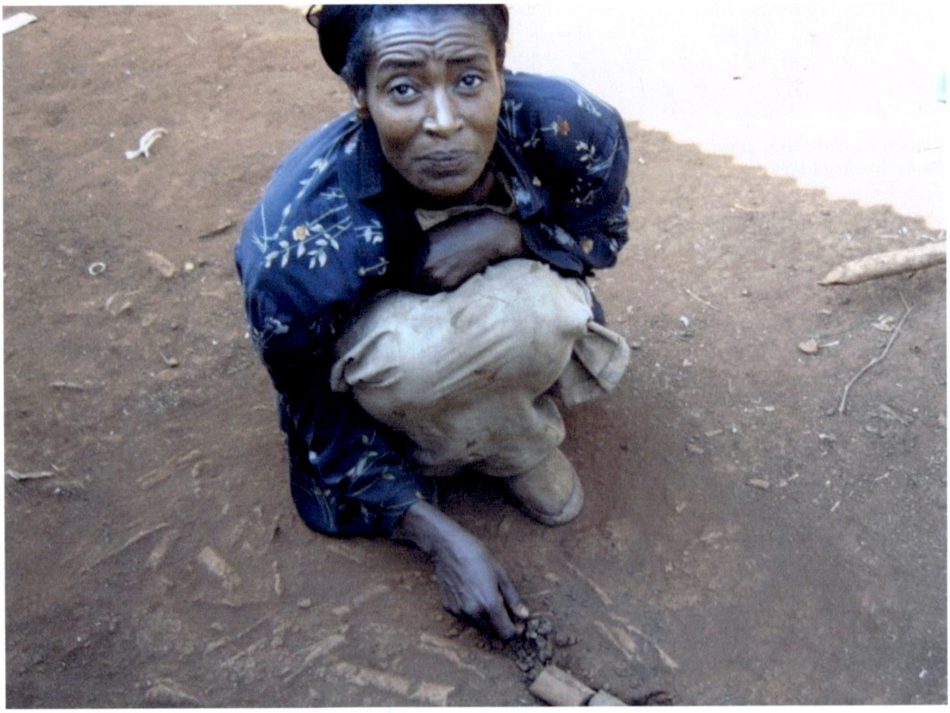

Figure 28 the photo above is taken from the compound of the woman in Aira town. The tuyères and fragments of slag are gradually exposed as she regularly cleaned in front of the door step.

smithing. One lesson from ethnoarchaeometallurgical reconstruction and oral histories is the fact that after each smelt, half way burnt up tuyères or broken ones were cleaned out of the furnace before the next smelting was ensuing.

In addition, it is not possible for a smithing site to create such debris of tuyères. It is known that smiths use only one clay tuyère and two bellows. The tuyère for smithing is short and cannot be longer than part of the tuyères, which are left out after a smelting is completed. According to smiths, it was common practice for their members to collect tuyère fragments for the forge and reuse it. They claim that such tuyères are much durable than they are making it afresh. These are macroscopic identification of iron production activities. It was based on the scenarios that survey anddocumentation of former iron smelting sites were conducted both in northeast and southwest Wollega. The following sites were then recorded on these grounds (see Figure 28 and the rest).

Like the preceding one, this site does not also belong to any one of the present occupants. Unlike the preceding one (Figure 28), however, this site (Figure 29) is located along a road track part of which is still extended to one of the house compounds beyond the wooden fence as it is clear from back side of the photo. This site is generally disturbed. Nevertheless, it is at further risk from future road construction and other development activities that the

Figure 29 this is another photo showing the upper surface of smelting debris in one corner of Aira town. it is exposed due to erosion and land degradation.

town is undergoing. As it is situated along a sloping land, erosion is another risk to this site. From my observation of the surface of the area,which is exposed due to erosion, (in particular after the upper soil was cleared during construction) the site could have been used over a long period of smelting activities.

In the spring of 2010, the municipality of Aira town has decided to use bulldozers as part of urban development program to widely opening up the intra-town roads. One of the new roads has cut through a former smelting site and revealed debris of vestiges of iron production activities (see Figure 30).The stratum consisting of vestiges of iron production was formed at about 40 cm and in some corners over 50 cm thick below the surface of the ground. Because of the slice by bulldozers blade that possibly took away part of the remains and dumped it elsewhere, a length of about five meters was exposed forming a distinct stratum from the rest. From this stratum, I am led to conclude that the smelting activity has been continuously conducted for some time before it was stopped. The fact that about 40 cm thick layer of soil was formed over the smelting debris suggests that its agecould not be a recent one. A big tree assumed under which smelting was conducted is located about 20 meters from the edge of the new road where the debris (slag and tuyères) were exposed. According to former smelters, the location of the smelting site could have been due to the need for a shade of the tree under which smelting house might have been built.

Figure 30 this photo is taken on 29/09/10 during the revisiting of formerly recorded smelting sites in the town. It was not previously recorded as no visible remains were observed.

Figure 31 this photo shows a block of slag fused together. When the undergrowth was cleared, a huge slag of about 80cm x 50 cm was exposed.

Part of the reason for conducting the 2010 fieldwork in southwest Wollega has been to fill in some of the gaps left unattended from previous fieldworks and build data for comparative purpose with ethnoarchaeometallurgical evidence. In previous fieldworks, the surveys were made along side with the reconstruction project of former smelting. An aspect of traditional metallurgy of the region, which was not very much understood, was about the separation of the impurities during the process of smelting. I was only informed that (e.g. Burka 2006, 2008) the bloom was dragged out of the furnace through the front bottom aperture. In this way, part of the slag was beaten off before it was further refined at the smithing. However, the remaining method of separation of the slag (e.g. Figure 31) namely the one that was formed at the base of the furnace was not clear. I was informed that the furnace was cleaned for the next round of smelting.

For instance, whether the slag was formed at the base of the furnace or a tapping method was devised to allow the flow of the impurities outside of the furnace was not very well ascertained. The ethnographic record we have observed during reconstruction in 2007 smelting could not allow us to develop enough interpretative models with regard to slag formation and separation. The identification of slag formation was not the major focus of the project. From the survey of former smelting sites in Aira town and outside of it (Figures 31 and 32), it was observed that a huge mass of slag was developed at smelting sites. Future investigation of this site would probably shed light on the level of technology attained for the separation of the impurities and the possible changes and continuities in this aspect. From the slag cakes discovered, it is possible to suggest that such a huge mass was formed in the space between the floor of the furnace and the tip-ends of the clay tuyères. In other words, Wollega smelters (particularly in southwest) have developed a special mechanism of separating the impurities and that when the bloom was retrieved very few slag could be left, which they used to knock-off during the first phase of refinement in front of the smelting house.

According to informants, the slag is exactly at a place where smelting was once conducted. The fact that the smelting left a mass block of slag estimated to be about 70 cm thick shows that smelting could have been installed here for many years (Figure 32). This conclusion is partly derived from ethno-histories that attribute to the permanent ownership of a site by certain family members and the tradition that offers that smelting site is a permanent installment. The challenge remains that in threereconstruction and re-enactment of iron smelting in both Wollega sites, no similar mode that results in slag formation was observed.

The first site with block of slag (Figure 31) was identified in a farming field at the edge of Aira town. The site is part of the corner of the town associated with iron smelting. Despite

Figure 32 one of smelting sites in Gujii Warabuu, about 5 km southwest of Gulliso town provides unusual size of block of slag mounds.

my intention to find explanation for this through test excavation, the result was completely different. The second site (Figure 32) where blocks of slag created a wall resting against a terrace believed to have been the place from where blowers worked on bellows was located about four/five kilometers south of Gulliso town. As a result, I would like to propose that in spite of ethno-historical testimonies of similar technologies, as far as memory was concerned, it was possible that various mechanisms of separating slag were adopted in southwest Wollega.

4.2.3. TEST PIT EXCAVATION IN A FORMER SMELTING SITES IN AIRA TOWN

One of the relatives of people whose history was associated with the traditional iron smelting practices was the family of Guutaa Sardaa (Figure 33). The significance of this site was that unlike other production sites, where people who do not have links with former smelters today occupy, the grandchildren of the last smelter, Guutaa Sardaa still own the land. Since the first survey of the sites in 2007, I established contact with the owners of the compound. The family head agreed that he would allow me for future investigation on his garden place where ancestral smelting site lies. He further assured me that the family did not have immediate plan for serious development activity except for sowing garden plants.

Figure 33 the granddaughter of the last smelter, Goobanee Olaanaa was sitting in front of the partly exposed smelting discard due to house construction that took part of it.

Figure 34 the test pit excavation was conducted inside the corn field. It means that the ploughshare disturbance was one of its characteristics.

Guutaa Sardaa's smelting site, which is now exposed due to human activities, is mainly inside the family compound and extends over 15 meters across the space but possibly extends to the neighboring compound to which I did not have access (Figures 33 & 34). My permission to have access to the site was facilitated by *Obbo* Oliiqaa Lamuu, who was integrated into the 2007 project for the reconstruction of smelting at Abayyanii (Burka 2008), about seven kilometers south of the town. He belongs to the family of the smelters, and has kinship ties with this family. Through his intermediary ship, I could negotiate with the family to open up a test pit (Figure 34).

I should discuss two contrasting issues about the Sardaa family's smelting site. On the one hand, the site is the only place in Aira town where members of the family who claim direct link to the site are still occupying it. Unfortunately, it is difficult to establish the nature of space distribution at the site level. In addition, no existing family members recall the exact location of activity areas.

Obviously, the objective of a test pit excavation is to determine the potential of the site for a subsequent comprehensive investigation. It is also important in that it gives additional information to those obtained through surface collection and investigation. In my case, surface finds mainly include broken tuyères and fragments of slag. However, a number of questions were still unattended, namely the depth of the exposed surface, the possible location of the foundation of the furnace and stone anvils upon which the first stage of bloom refinement was conducted as well as the proximal section of the clay tuyères.

With these and aim of testing the potentiality of the site for future comprehensive investigation, I decided to open up a pit. My choice of the site for the test pit was arbitrary. However, it draws experience from ethnoarchaeometallurgical evidence. From the ethnoarchaeometallurgical evidence, we knew that smelting was conducted from upper slop downwards. As a result, I opened a 1m x2m pit in the middle of the compound and mound with the help of three laborers. I also decided to go as deep as the vestiges were accumulated (see Figure 34). The site is a disturbed one as it has been under cultivation for long for growing garden plants.

Accordingly, below the undergrowth, which was cleaned from the intended area of the test pit, the soil can be characterized as black humus soil and of burnt nature interspersed with metallurgical remains. The conspicuous components of the remains were tuyère fragments of about 12 cm in length but generally vary. Interestingly, the conspicuous components of tuyère fragments belong to its distal parts as evidenced from their slagging tips. There were also the mid-sections of the tuyères without any slag formation on either of the sides. These findings fit with previous surface collections at different former smelting sites. Of few of the finds was the proximal part of the tuyèe (the head) measured to be about five centimeters (Figure 35left) and a metal rod badly abraded (Figure 35right).

An important contribution of the shovel test was the retrieving of the upper part of the clay tuyères. It has never been recorded during previous surveys where surface collections were closely examined. In other words, the finding provided not only an evidence to fill in the gaps created during previous field survey records but also raised further questions due to the absence of substantial size of slag as exposed and recorded from other sites mentioned above. The shape of the upper section of the clay tuyères, on the other hand, was only known from ethnographic records during reconstructed and re-enacted smelting activities. In this procedure, the excavation has continued to the depth of 60 cm from the

Figure 35 the proximal section of the clay tuyere, madabii (left) and a metal rod probably locally made nail of about 10 cm long(right), which is still magnetic

surface of the ground. While it appeared that the conspicuous finds (tuyère fragments) would continue to dominate the remains, it however gradually decreased in quantity. Then at the depth of about 60 cm, the color of the soil abruptly changed from dark and burnt ashes to a reddish color probably suggesting that the depth of the metallurgical culture was over.

In sum, the archaeological survey of sites associated with iron production in southwest and northeast Wollega study areas expanded our knowledge of the nature of the activity in the region. The archaeological survey has confirmed in many regards local histories and ethnoarchaeometallurgical evidence on the spread and nature of mining sites, smelting areas, and smithing sites. In addition, it has also shown the possible relevance of cultural preferences as opposed to environmental factors on various human decisions in their interaction with the surrounding environment. Moreover, it has contributed in distinguishing features peculiar to each of the two study areas. For instance, in northeast Wollega of Abee-Dongoro study area, smelters and smiths had limited options with regard to procurement of iron ore.

The survey has attested the local information that there was no mining site in the vicinity other than that of the Faallee area. In addition, whereas iron-smelting sites are identified across tens of kilometers in southwest Wollega of Aira and Gulliso districts, in Abee-Dongoro there is only one village associated with iron smelting. It shows that control of knowledge of iron smelting in Abee-Dongoro was more structured compared to the one in the southwest.

The centralized clan control of mining sites and the knowledge of smelting might explain for its limited spread across the space in the district and its environs. In contrast, in southwest Wollega, the absence of local lords whose history attributes them to also own smelting sites might have been the reason for the wide spread of its activities throughout the region. In addition, the number of artisan groups in southwest Wollega was much bigger than the ones confined to a single *ganda* Walagee in Abee-Dongoroo district. Generally, survey of archaeometallurgical sites in Wollega emanated from the conviction that a good picture of the whole of things is made possible when we know how different parts operate their own respective ways.

CHAPTER FIVE

5.0. THE ETHNOARCHAEOMETALLURGICAL RECORD: CASE STUDIES FROM WOLLEGA

5.1. INTRODUCTION

> Anthropologists do not assume that intensive investigation of each individual group will produce a valid picture of the whole. A more likely belief is that no valid picture of the whole can emerge if we know nothing of the parts. (Todd 1978:146)

It has long been noted that the historical picture of the archaeology of iron production in Ethiopia is incomplete (see Burka 2006, 2008, 2009; Haaland 1985, 2004a; Phillipson 1993, 2005; Todd 1985; Todd and Charles 1978a, 1978b). The lack of complete picture of archaeometallurgical history of Ethiopia is partly due to the focus of Ethiopian archaeology on elite sites including mainly monumental architectural sites and exotic goods (cf. *Chapter I, VII*). The little available evidence of metallurgical history is only found from sporadic ethnographic documentation of 'Late Iron Age' (ethnographic) sites in various parts of the country.

In an attempt to redress these gaps, this chapter seeks to provide a detailed coverage from ethnoarchaeometallurgical sources and as a result forms the core chapters of the dissertation. In terms of detail of a holistic approach and work by an insider, the ethnoarchaeology of iron production in Wollega is the first of its kind in the archaeology of iron production in Ethiopia. Thus, this chapter provides a comprehensive perspective on the technology and its socio-cultural aspect. Previous scholars have only tangentially addressed these issues (cf. *Chapter I* and below).

Furthermore, the fact that the native of the region conducts this research is particularly significant in redressing the gaps left hitherto in the scanty documentation attempts by expatriate scholars. The previous attempts lack the scale and comprehensiveness attained through the current successive researches that resulted in this dissertation (see Burka 2006; Cline 1937; Haberland 1959; Haaland 2004b; Haaland *et al* 2000; Haaland *et al* 2004; Huntingford 1955; Lewis 1965;Todd 1985; Todd and Charles 1978a, 1978b; Wainwright 1942a, 1942b). In this chapter, I present the ethnoarchaeometallurgical case studies from Wollega in line with the theoretical and methodological frameworks I have outlined and explicated under previous chapters (cf. *Chapter I and III*). The technological variables I have outlined under *Chapter III* in line with the *chaîne opératoire* theoretical framework are further developed here through ethnoarchaeometallurgical data from the study areas.

This chapter is divided into two parts. In the first part, I discuss previous sporadic ethnographic studies (*cf. chapter I*) carried out on iron metallurgy in central and southwestern Ethiopia (e.g. Cline 1937; Haaland 2004a; Haberland 1959; Huntingford

1955; Lewis 1965; Todd 1985). Lack of ethnographic record on the subject for most of northern Ethiopia has limited the opportunity to compare archaeological materials derived from Wollega's immediate neighbors around the Blue Nile. The first part of this chapter is, therefore, an extended discussion of issues, which I have begun to outline under chapter one. After providing an account of the previous attempts made to document indigenous (traditional) iron smelting practices, I proceed to the second part of the chapter. As part of introductory remarks to this part, I will also outline some scenarios relating to the possible significance of iron metallurgy in northern half of the country since the antiquity.

In the second half of the chapter, the discussion focuses on areas selected for this study in Wollega, western Ethiopia. Under this section, I present the two case study sites of Abee-Dongoro and Aira-Gulliso situated in the northeastern and southwestern parts of Wolega[34] in western Ethiopia. As I have already noted (*Chapter II*), the study sites are located in the districts of Abee-Dongoro and Aira-Gulliso in the east and west Wollega administrative zones respectively of the Oromia Regional State. In this section, I structure and put together my observation and documentation of a reconstruction and re-enactment of traditional iron smelting practices, local tradition, interviews, and survey results.

The discussion is the sum total of ethnoarchaeometallurgical and ethno-historical research initiated in northeast Wollega in 2005 and then expanded in 2007 to include sites in the southwest and other parts of the region to the year 2010. The chapter deals with the study of the smelting technology and its socio-cultural aspects and owners of the knowledge.[35]

Various archaeologists working on fields other than archaeometallurgy in the Ethiopian region and the Horn have long noted that Ethiopian craft archaeology lacked focus on the study of the technological tools that contributed to the development of ancient culture. It also lacked focus of study on the perspectives of people associated with craft technology, whose position is regarded low by the mainstream section of the society. Generally, Ethiopian archaeology is still suffering from spatial and topical imbalances.

Wollega is one of the most underrepresented regions in the country with regard to Ethiopian regions under archaeological projects.[36] The topical imbalance is particularly glaring in the archaeology of iron production, which, of course, can be said also of the rest of the Ethiopian regions. Similarly, Ethiopian archaeology left, in the backwaters, the role of metal

34 It must be understood that the mere division of Wollega into different zones possess no cultural connotation other than an attempt for administrative convenience. Despite this, however, separate administrative histories are expected to create variations in some of cultural manifestations. For instance, northeast Wollega has been dominantly Orthodox Christians as opposed to the major Protestant influence in the southwest.

35 Thus, there is no a priori reason to argue that the current ethnoarchaeological research provides a complete picture of the activities as conducted during the active operations in the past. However, an archaeological enquiry inspired with data from ethnoarchaology is well-off than the one who conjures on the information derived from 'mute' objects.

36 Except for my ethnoarchaeometallurgical projects and a recent project in the western lowlands of Ethiopia by Spanish scholars, there has never been any attempt by the ARCCH, the institution in charge of the national archaeological research, to include Wollega in its archaeological inquiry and conservation of cultural heritage of its peoples.

tools in various monumental architectural developments in various parts of the country since antiquity. Ironically, these monumental features are the major focus of Ethiopian archaeology since its formal beginning by the turn of the 20th century. As agreed by many scholars (see Haaland 2004b; Mapunda 1997; Phillipson 1993; Todd 1985) and mentioned earlier in this work, only indirect remarks are available particularly for the early phase of iron production in Ethiopia and the Horn.

From the existing patchy literary and archaeological evidence, one can conclude that metallurgy has become part of other cultural traits of the peoples of the Ethiopian region and the Horn as far back as the middle of the first millennium BC or even earlier (e.g. Anfray 1981; Chittick 1978; Contenson 1981; Phillipson 1977, 1985, 1993; Wainwright 1942a, 1942b, 1942c). The development of sophisticated stele erecting might not have been concurrent with seems the appearance of metals in the archaeological contexts of the northern Ethiopian highland regions (e.g. Aksum). However, the close location of the stele quarry site and the Gobedra rock shelter (see Phillipson 1977) where remains of significant iron working of early first millennium BC were discovered, though not necessarily causative, might not have been coincidental.

It is clear that Aksumite steles (see Figure 36 middle) represent and symbolize anthropomorphic shapes. The main stele in particular has a head with pronounced male's glandular genital. Anintriguing finding of its metal prototype is bronze bracelets shaped with male's genital ends on both sides, which have been discovered close to the main stele field (Figure 36 left). One may ask whether this too is accidental. Given the pan-human ritual perspective towards metals in general and iron in particular, it is worth mentioning an ethnographic ark-like object, the Oromo's *Kallacha* (Figure 36right), in the possession of the ritual heads, the *Qaalluu* (Burka 2009) for comparative purposes.

Previously (cf. *Chapter I*), I have argued that if it were not for the discovery of finger-millet's intrusiveness that disturbed the credibility of the dating at Gobedra rock shelter (Phillipson 1993; Haaland 2004a), iron metallurgy remains found in the early phases of

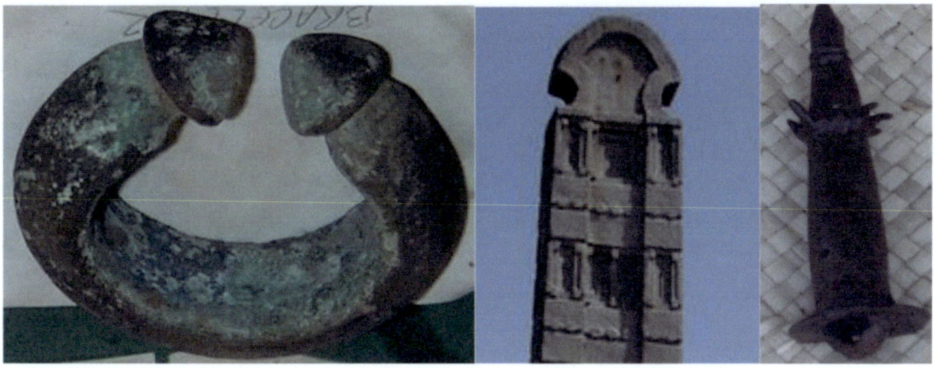

Figure 36 bronze bracelet from Geza Agumai (left) close to the main stele field and the head part of one of the major stele (middle) in Aksum and the Oromo kallacha (right) in Wollega (Butaa Bulluq ritual (Qaalluu) site)

the site simply suggest the antiquity of metallurgy in Ethiopia. If this were the case, one can posit the likelihood of the spread of the knowledge to different parts of the sub-region because of population movements and contacts, and exchange of goods (see Todd and Charles 1978a). On the other hand, the association between phallic bracelet and the stele with emphasized genital head may indicate the cultural embeddedness of the technology. The carving of metal with phallic heads symbolizing man's strength and fertility is also available elsewhere as in the Oromo's *Kallacha*.

Undeniably, the Aksumite Empire whose southern extent is not well defined possessed an advanced technology of metallurgy such as in minting coins (e.g. Munro-Hay 1991, 1993). The questions that remain unattended include- what happened to the artisans following the collapse of the empire? Did they also accompany the royalty in the so-called southward 'retreat' (see Sellasie 1972; Tamrat 1972)? After all, of the research that focused on elite site in Aksum, it was not possible to locate where the seat of the empire was shifted after the collapse of the empire.

Lack of research on metallurgy is also evidenced from the medieval Agaw kingdom of the Zagwe in north central Ethiopia, which is mainly represented by its highly refined rock-hewn churches of Lalibela (Selassie 1972; Taddesse 1972). The use of pickaxes and chisels for carving the monolithic churches from rock masses of volcanic hills is understood. Nevertheless, particularly metal doorframes and decorative nails on the doors of the churches are good evidence of the abundance of metals in everyday use of the society (Figure 37). At least, one can posit that metallurgy has achieved high level of development in the kingdom's project of carving out the monolithic churches. Furthermore, a comparison of decorations of one of the incense[37] burners from the *Qaalluu* site in Wollega (at *Irreessa* ritual site) where a similar decoration in metals is available suggests the need for detailed inquiry in the future.

The 16th century is often considered a decisive landmark in Ethiopian culture-history (e.g. see Gidada 2001; Hassen 1990; Ta'a 1980; Tamrat 1972). The so-called Oromo population movement and 'occupation of new lands' in northwestern Ethiopia is considered to have changed the population picture in the later history of Ethiopia. Some argued that the Oromo were 'nomadic hordes' that destabilized the 'civilized Abyssinian' empire and contributed nothing to its civilization (see Bruce 1773; Hassen 1990; Jalata 1996; Megerssa 1994; Ullendorff 1960). However, one still can posit that what came to be known as the Gonderine kingdom (see Abir 1969; Hassen 1990; Zewde 1991) has resulted from combination of ideas and cultures. The fact that one of its early founders by the name of Susenios (r. 1608-32) assumed power with the help of the Oromo gives the weight of their role in various developments of the kingdom (see Hassen 1990). The presence of Oromo influence on the Gonderine Kingdom through Susenios is found in that he was an adopted Oromo himself.

37 Geographically speaking, Wollega and northern Wollo are untenable for comparison. However, the decorative techniques of the incense burners, as in the figure, and other ritual objects such brewing barrels pose similarity with the decorative techniques of doors at Lalibela though this was used with iron nails. That iron possesses significant ritual place both in the traditional Oromo religion and the Ethiopian Orthodox Church also adds weight to our discussion about the materials.

Figure 37 decorative nails on one of the doors of Lalibela rock hewn churches (left) and incense burner at traditional Oromo ritual site in northeast Wollega

On cultural grounds such as the knowledge of metallurgy, the period could have been a suitable time for diffusion and adaptation from one part of the Ethiopian region to the other. The smithsmight have accompanied the Oromo as far as Gondar or such population dynamics in the region particularly facilitated by the two Oromo institutions of adoption (cf. *chapter II*) might have allowed free movement of itinerant smiths to disseminate their knowledge. Unfortunately, lack of record on the type of technology north of the Blue Nile (in both Gojjam and Gondar) restricts our imagination towards regional comparison.

Later in the 19th century, the controversial king Teowodros II of Ethiopia is best known for establishing a gun foundry at Gafat in southern Gondar (see Zewde 1991).The historical discourse of the period is caught up with the emperor's imprisonment of British nationals that eventually caused the loss of his life. The discourse also focused on the emperor's forcing of forging the arsenal. However, as regards the establishment of the gun foundry at Gafat, one must underline that the decision to select the site could have been embedded in the king's knowledge of the local cultural roots of metallurgy. From these patchy examples in northern Ethiopia, it is very clear that there is not enough data to allow us to discuss the archaeology of metallurgy all through prehistoric and historic periods.

Nevertheless, it is clear that any archaeologist doing research on a collection of material remains (debris of past activity areas) seeks to reconstruct the actions and processes that led to those remains. In the case of early iron metallurgy sites, for instance, only few production debris (e.g. slag, tuyères and furnace fragments) are available providing the archaeologist with a less clear picture of how production was conducted. Other issues such as how producers structured their decision or what decisions they took, and the nature of environmental and cultural constraints are always too complicated for the archaeologist to reconstruct from fragmentary remains.

At this stage, alternative approaches such as ethnoarchaeometallurgical ones are sought to help generate models upon which the interpretation of the prehistoric past rests. Few of these works are, however, available for Ethiopian metallurgy. Part of the reason arises from the absence of the continuity of the tradition in various parts of the Ethiopian region and the Horn. The other reason could be found in lack of research focus on this particular theme in Ethiopian archaeology. In the following section, before I discuss my case studies in Wollega, I first summarize some of the previous observations and descriptions of traditional iron production in different places in Ethiopia and the Horn and periods to allow us obtain a better comparative perspective.

5.2. INSIGHTS FROM ETHNOGRAPHIC RECORDS IN CENTRAL AND SOUTHWESTERN ETHIOPIA

Remarks on iron production and iron as one of the trade items in the long distance caravan trades in the Ethiopian region and its neighbors go as far back as the 6th century AD (Wainwright 1942b). Walter Cline (1937) is one of the earliest workers in our period to remark on aspects of iron working technology (furnace, charging mechanism, and bellows) among the Oromo in Shoa but without specific sites. His table of attributes of iron smelting among the Shoa Oromo is related to the size of the furnace, the operation mechanism of the smelt, its location vis-à-vis resource procural sites and the decision of the smelters after the work is over. Wainwright (1942a) on his part made comparisons of terminologies used for iron by different ethnic groups in the Sudan, Eritrea, and Ethiopia such as the Oromo. His reference to the Oromo term of *Sibiila Gurraacha* for iron is significant in that the term is often associated with the myth of its origin (Burka 2006, 2008, 2009, 2010).

Another discussion on iron working among the Oromo is found in Huntingford (1955) and Lewis (1965). Their work in the southwestern part of Ethiopia in the kingdom of Jimma Abba Jifar, one of the five Oromo Gibe states of the 19th century (see Abir 1969; Hassen 1990), provides a different picture on the subject. It shows in particular that iron production was under centralized control of the palace. Eike Haberland (1963) who wrote about the Oromo of southern Ethiopia has ruled out the presence of the culture of metallurgy among the Oromo of southern Ethiopia and asserted its outside origin (see Todd 1985).

Haberland's presentation is criticized for its shortsightedness particularly for extrapolating his 'unreliable' information from southern Ethiopia and for presenting it as if it were representative of the Oromo in general (Burka 2006). Unfortunately, his work coincides with the period when palace clerics and their European cohorts such as Ullendorff (1960) who defined the Oromo as 'destructive' forces and 'without material culture' were creating fictitious stories about the Oromo (Burka 2006; Megerssa 1994).

Haberland on the other hand (1959 in Haaland 2004a) offers an early eyewitness account of iron smelting activities in the Omo valley. His description was based on his observation of iron smelting carried out by the inhabitants of the Omo valley. This early phase of observation and report of traditional iron smelting in southwestern Ethiopia led to two separate studies on the technology and socio-cultural aspects of the craft. Judith Todd conducted the first of these in the mid-1970s (see Todd 1985; Todd and Charles 1978a; Todd and Charles 1978b). She carried out her study among the Omotic-speaking Dime community of Southwest Ethiopia (Burka 2006; Mapunda 1997; Todd 1985). Todd is a material scientist not an archaeologist, however.

As a result, her work on the Dime was undertaken from the material sciences or technological point of view (Burka 2006; Haaland 2004b). It lacks what can be called an ethnoarchaeometallurgical perspective as it fails to address the cultural aspects of metallurgy. This is clear from her emphasis on the technological aspects for tracing origins (e.g. Todd 1985; Todd and Charles 1978a, 1978b). She explained that the smelting was the last remnant of the knowledge in the region (see Haaland *et al* 2000; Todd 1985). However, after about 30 years, it was still found to have continued to allow a further research on the neighboring communities of Oska Dencha in Dawro (Haaland *et al*, 2000, 2004a, 2004b) where the activity continued as part of the socio-economic life of the inhabitants.

A team consisting of an archaeologist and anthropologists from the University of Bergen, Norway carried out an iron smelting research in ethnographic setting among the Tsara (also Omotic-speaking) neighbors of the Dime in southwest Ethiopia (Haaland *et al* 2000, 2004a, 2004b). It involved a two-season observation and documentation including filming of the procedures. Although the study addressed both technological and cultural aspects of the metallurgy, it lacked comprehensiveness.

Although the team could distinguish between the smelters (belonging to potters) and smiths (as a separate identity), they did not present the reason on how the master smelter could work as a smith to produce tools from the bloom. (Was he a smelter who settled among smiths and learned from his hosts? Was he a smith who obtained the knowledge as an apprentice?) According to the team, the technology of iron smelting (explained through furnace, tuyères, and bellows) was much similar to what Todd had previously described for the Dime people. The reason for this similarity, according to Randi Haaland (2004b), was that the smelter at Oska Dencha had migrated from the Dime country.

In both areas, the smelting apparatus is characterized in the following ways: shaft furnace erected on a slag pit, forced draught, drum bellows and (or generally) non-slag tapping smelting technology. Yet, there are minor differences in sizes and thickness of the furnace and the number of tuyères. For example, the Dime furnace was about one meter tall surrounded with six clay bellows connected to 36 tuyères (Todd 1985). The furnace superstructure at Oska Dencha was about 80 cm above the ground and operated with five clay bellows connected to 30 tuyères. In both sites, the bloom was retrieved through the top opening of the furnace after it cooled down and the furnace was not then destroyed (Haaland 2004b). The length of the tuyères is about quarter of a meter. The works at Dime and Oska Dencha represent the culture of Omotic societies. How is the Omotic smelting culture comparable with the Oromo smelting technology?

Halland (2004b) discussed the Oska Dencha smelting in the context of hierarchical society sanctioned by divine kingship. Both Herbert Lewis (1965) in his ethnographic work among the Macha Oromo of Jimma in the late 1950s and early 1960s and Huntingford (1955) explained some features of iron working technology by people who were directly responsible to the agent of the king. From the geographical point of view, the kingdom of

Jimma was not only close to the Omotic speaking peoples but also within the Gibe river system.

The kingdom of Jimma is said to have originated in part because of its strategic location along the trade routes that connected the Omotic inhabited region of Southwestern Ethiopia with the central and northern regions through which a number of trade items including slaves circulated (Abir 1969; Hassen 1990). The kingdom of Jimma was an Islamic kingdom that had international connections through various forms of exchange networks (Burka 2004).

Like the smelters of Dime and Oska Dencha villages of the lower Omo valley, Oromo smelters of Jimma in the upper valley of the river used a forced draught furnace. The difference is particularly in size where the Jimma furnace was about two meters tall while that of Dime and Oska Dencha was not over one meter (Burka 2006; Haaland *et al* 2004a, 2004b; Huntingford 1955; Lewis 1965; Todd 1985). In addition, both differ in terms of the kind of bellows smelters employed, namely pot bellows for the Omo valley and bag bellows for the Jimma ones. Smelters of the Jimma kingdom seemed to have shared this technological feature with other Oromo smelters in different parts of the region than their closer neighbors.

In spite of the geographical proximity of both iron production sites under investigation in the Gibe-Omo valley, there is marked difference in the use of technological facilities. For instance, among the Omotic smelters, the number of tuyères used is over 30 as opposed to one or two mentioned for Jimma (Huntingford 1955; Lewis 1965). Obviously, technological variation in iron smelting is widespread even among the Oromo smelters in various regions. Technological variability is evident from patchy records in different parts of Oromo inhabited regions of Ethiopia.

In this case, despite it is posited that the Oromo population had a common origin (see Hassen 1990), a remarkable difference in their smelting technologies in different geographical or environmental settings is evident (Burka 2006, 2008). A glimpse of comparison of sites in central Ethiopia (Shoa), western regions (Wollega) and southwestern areas (Jimma) among the Oromo inhabited areas shows the validity of this argument. For instance, the smelters of central Ethiopia in Shoa (Cline 1937), erected a large furnace of about two meters tall and operated it naturally and destroyed it after each smelt.

Cline (1937), however, adds that there is occasional kindling of the operations to force air into the furnace. He is vague, however, on the level of involvement by the smelters during the process. What does occasional kindling mean? What part of the furnace is designed for this occasion? What types of bellows were employed? What was the level of destruction of the furnace? Was the furnace totally or partially destroyed? Such questions remain unanswered since no other data is available for the site to, at least, corroborating it.

The description for both Jimma and Shoa furnaces is almost similar. In both areas, the scholars have mentioned that iron smelting was conducted close to mining sites. What does it mean by stating that the furnace was built at the mines? On the other hand, Jimma smelters employed a forced draught mechanism operated with one or two tuyères (Huntingford 1955; Lewis 1965). What does it mean by one or two tuyères? Does it mean to distinguish between smelting and smithing activities as in Wollega (Burka 2006, 2008)? Like the furnace of Shoa, the description for Jimma (Huntingford 1955; Lewis 1965) suffers from various shortcomings. For instance, it is not clear whether smelters and smiths had different status. It only indicates that smelting and smithing were physically separated. It also lacks clear indication on whether destruction of furnaces after smelt refers to a portion of the furnace or total demolishing of the furnace (e.g. Burka 2008). Another important question is also, whether iron ore was transported over long distances across the region outside the control of the king. It is only mentioned that both smelters and smiths were under central control of the king or king's representative oversaw them.

Cline (1937), Huntingford (1955) and Lewis (1965) referred to iron working in areas occupied by the Macha and Tuulama Oromo moiety (cf. *chapter II*). One common feature of iron smelting explained by these early scholars is the fact that the Oromo smelters used bag bellows and their furnaces were all with superstructure. On the other hand, although Wollega is located on the triangular edge of Shoa and Jimma in western Ethiopia inhabited by the Macha Oromo (Gidada 2001; Hassen 1990; Hultin 1984; Knutsen 1984; Ta'a 1980; Triulzi 1996; Waqweyya 1988) there is no ethnographic observation and record for iron smelting in its different parts before it stopped.

The presence of iron production activities in different sites in Wollega can be gleaned from passing remarks made in the historical literature (e.g. Ahamd 1980, 1999; Triulzi 1973, 1975; Waqweyya 1988) on the circulation of the products. In his study of trade and politics in Gojjam, Ahmad (1980) argues that the Gojjam region (north of the Blue Nile) has mainly depended for iron and its products from Wollega. Similarly, quoting 19th century travelers who visited markets, Triulzi (1973) mentions iron amongst items of trade sold to merchants north of the river at market places on the boundary of the two regions(see also Kenno 2007). It was referred to as 'iron of Horro'.

Horro is a highland region in northeastern Wollega, which provides multiple tributaries to the Blue Nile (overlooking the Abe-Dongoro lowlands in the west) and known for its variety of crops (e.g. barley, wheat, *teff, noog* peas and beans, etc) and the Zebu and sheep breed (Burka 2006; Waqweyya 1988). It is also the center of traditional Oromo religio-secular system evidenced from the site at Gadaa Bulluq (see Burka 2009). The Oromo ritual head, the *Qaalluu* of Butaa Nadoo at Bulluq provides the role of resolving conflicts in the medium of metal tools. The source of iron tools for the ritual sites as well as secular purposes in the region used to be obtained from lowland Abee-Dongoro (Burka 2009).

However, it was only from other historical literature that indigenous iron smelting localities could be identified (see Waqweyya 1988). According to Chernet Waqweyya (1988), the location of indigenous iron producing area in northeastern Wollega was at a locality called Walagee in the district of Abee-Dongoroo. He did not, however, go further and discuss the type of the technology the artisans used. In addition, he did not have details on whether those who carried the knowledge were still alive or not. His mention of iron production came as part of his general discussion on the socio-economic significance of the region and as a center of contest between local lords in the 19th and 20th centuries as well as newcomers into the region from the late 1960s.

The significance of iron as part of trade item from southwestern Wollega is also found in historical literature provided as passing remarks. According to Bahru Zewde (1987), merchants from Wollega exchanged their iron for cotton from Gambella, among many other items. The other significant source is the source by Alessandro Triulzi (1975, 1981).

During his research on population contacts in western Ethiopian regions, Triulzi identifies that the highland regions of western Wollega provided items such as durra, iron, and others for exchange not only within Ethiopia's national boundary but also across the border to market centers such as Karakoj, Roseriy and Fazoqli in the Sudan. Other scholars also noted the penetration of Arab merchants during the 19th century from the Sudan across Wollega into the Gibe region of southwestern Ethiopia (Ahmad 1999; Hassen 1990). Does this allow us to infer that Wollega has been serving as the source of various products to the outside world since remote past? Back in the 6th century AD, the region around Wollega (e.g. Sasu=Asosa?) has been the source of gold for the kingdom of Aksum (Wainwright 1942b). This shows that the 19th century was not the first time when merchants penetrated the western Ethiopian region from the Sudanese side. As Wainwright (1954 in Todd and Charles 1978a) argues, the fact that Wollega might have been part of the southeastern farthest extension of Meroitic influence provides a ground for suggesting that there might have been various ways of acquiring ideas and objects from outside and vice-versa.

It is clear from these early works (Ahmad 1980, 1999; Hassen 1990; Triulzi 1973, 1975 and 1981; Waqweya 1988; Zewde 1987) that Wollega is as an important region that has served as a bridge connecting peoples across regions. The bridging role of Wollega was between southwestern Ethiopia beyond the Gibe River (e.g. Kingdom of Jimma) with northern Ethiopia across the Blue Nile (Gojjam) and the Sudan across the western lowland Nilo-Saharan speaking population. However, the most important point to be underscored is the fact that Wollega has also been a source of a number of products such as iron and cattle, which were exchanged within different parts of the Ethiopian region and beyond its national boundary.

5.3. ETHNOARCHAEOMETALLURGY IN WOLLEGA: REVITILIZING INDIGENOUS KNOWELDGE

In the following section, I present my findings through ethnoarchaeological enquiry of smelting activities in northeastern and southwestern districts of Abee-Dongoro and Aira-Gulliso. The discussion under this section is the result of the following activities. First, it emanated from a 2005/06 re-enactment study I have conducted among the Malkee/Yamalagii smelter/smiths of Abee-Dongoro in northeast Wollega. Secondly, a similar research was initiated in 2007 in which I have reconstructed former iron smelting practice among Laaloo/ Abayyani smelter/ smiths of Aira-Gulliso district in southwest Wollega. Finally, between 2008 and 2010, I have conducted various field works for expanding my perspective on different aspects of the life of the society (e.g. mainstream) including the smiths/smelters as part of this PhD study.[38]

38 Since 2008, four fieldworks have also been conducted to augment the information hitherto I have regarding metallurgy and the people in the study area. This chapter presents the activities involved in the reenactment and reconstruction in the summer and autumn of 2005 and in the late 2007. However, as discussed under the previous chapter, other fieldworks (in 2009 and 2010) were conducted in order partly to retrieve data for enriching those obtained during ethnoarchaeological reenactment and reconstruction projects. Particularly, the rituality of iron (which will come in the subsequent chapter) that we have observed in connection with the *chaîne opératoire* of iron metallurgy needed particular fieldwork around traditional ritual areas.

Although the location of the two sites is separated by hundreds of kilometers, due to close links between the artisans of the regions, they share a number of common traits. As a result, particularly regarding the making of tuyères, and bellows, I carry out one discussion representing both sites while I add some of the varieties that distinguish one from the other. I will distinguish between the two sites with regard to the techniques and procedures they used for ore (iron and clay) processing and charcoal making. The description concerning furnace also varies due to the presence of significant technological divergence between the two sites (Burka 2006, 2008).

In accordance with the methodological framework of the *chaîne opératoire,* I have introduced under chapter III, the archaeology of iron production is best understood through its trajectory of production, use, reuse, and discard. In this context, the *chaîne opératoire* concept becomes very meaningful, explanatory, and insightful. In the following section, I start with the description of the variables and the procedures that lead through production processes in transforming the natural raw material to cultural objects. This also offers a significant vantage point to observe how people respond to the challenges and prospects they face from both human and natural environments at every juncture of operational sequences.

5.3.1. *GORDANA BAASUU AND GORDANA MURUU*: IRON ORE PROCUREMENT, PREPARATION, AND USE OF ABEE-DONGORO AND AIRA-GULLISO

The iron ore-mining site of Faallee in Abee-Dongoro district is said to have been in the territory controlled by the Malkee chieftaincy (see *chapter II*). In contrast, although Malkee clan members inhabit part of Aira-Gulliso district, the chiefs from the Laaloo clan controlled the ore-mining site at Wiinsaa Kacho. Unlike Abee-Dongoro smelters who were limited to one site at Faallee (Burka 2006), Aira-Gulliso smelters had more than one site to procure ore (Burka 2008; cf. chapter IV).

Although at different times, particularly in the later period, people from different groups conducted smelting, the smelters mainly belonged to the Malkee clan[39] in the northeast and Laaloo (as well as Malkee) in the southwest (Burka 2006, 2008). In both areas, the local Oromo term for iron is *Gordana*; hence *gordana* Faallee, *gordana* Wiinsaa Kachoo, *gordana* Dimma. Whereas miners in southwest call iron mining *gordana muruu,* the procedure in the northeast is known as *gordana baasuu*. Some practical difference can be discerned in the two terminologies. The former literally connotes 'to cut ore' while the latter means 'to take out ore'. In both areas, the dominant ore is hematite (*Chapter II*) but differ in terms of physical properties namely powdered vs. solid rock.

39 It must be noted that whether in the northeast or southwest, smelting or smithing whole members of respective clans have never conducted. However, clans with suffix *inna* or *hinna* as in *Innagamdii, Innanaanii, Innabalxii* are generally regarded as craft workers. This ascription is found in the myth of origin of the smith and the *Qaalluu,* which I have presented in other sections of the dissertation.

Therefore, the difference in terming ore mining by the locals might have been rooted in the sizes of the ore and nature of mining. In the northeast, they retrieved it in a powdered form as opposed to cutting it from blocks in the southwest. In some other economic activities such as honey collecting from beehives in the northeast, the process is known as *damma baasuu*, literally 'to take out honey'.

The mining pits in both localities are markedly different. As became evident from surviving pits, the holes are narrow to allow no more than one person to enter at a time as opposed to the widely cut circular or rectangular or even irregular pits in the southwest (Figure 38). Traditions hold that only two persons used to enter to mine and to send up the matrices, which assistants pulled it up by leather strips tied to wooden bucket.

According to old miners of Abee-Dongoro, the procedure involved cutting the ground in a circular form and digging deep until they arrived at what they called *hidda gordanaa*, meaning the ore vein. No one volunteered during the survey to enter into one of the pits to prove the statements due to a number of reasons. First, the mining site has become a ritual site for members of the Malkee clan over which ancestral spirits particularly the woman founder hover and revered. Second, they believed that the pits host dangerous beasts such as pythons, which are revered as well as feared. Third, they also noted that the pits, due to long age ages, became fragile and can collapse on them (*cf. chapter IV)*.

Obbo Moosisaa Gaaggaa, an old miner from senior member of Malkee lineage stated that the discovery of the vein can happen at about 10 meters of depth, but in most cases, veins go up to 20 meters in depth. He also said, and showed to us some shallow pits, that people who had out of greed dug in front of their gardens to take out the ores found nothing. He ascribes this to special confinement of the ore to heavily mined site where the legendary woman, Balfee Saqoo,[40] discovered it. He in addition attributes this to the belief that supernatural forces are responsible in revealing or hiding ore to people. Through this, he justified why his close lineage were in charge of the mining and smelting for which lineage members made annual sacrifice of an old cow (*dullacha*) to the site was until the mid-1970s.

According to the old miners, once the vein was identified mining was conducted horizontally following its line.[41] Cutting sideways into the banks is also said to have been practiced by southwestern miners. Another old participant in illicit mining in southwest Wollega stated that while he was mining with his friend, part of the opening collapsed on his friend, who the surrounding people subsequently rescued. In the mean time, he confessed that he was not experienced miner and he went to the site hiding from his father and the landowners and the purpose was to get some money for some expenses. This is evidenced that smelters used to obtain ore from different sources and that ore had a significant market in the region.

---

40 Balfee was the wife of one member of the Malkee clan. She was a granary worker when she discovered the ore. It was unusual in Oromo to ascribe the discovery to a woman because in the past land belonged to husbands. The name Balfee is also that of a notorious spirit that causes epidemic diseases.

41 Miners of different pits on occasions met underground and created disputes over 'boundaries'. If they could not settle them it there, they used to take to the court at the local chief of the clan. Another incident they remember two miners were the pit of because a bull fell half into it blocking them an exit. They had to wait until members of the family worried for their unusual lateness came and relieved them.

Figure 38 one of mining pits at Faallee (above) and Wiinsaa Kacho (below) in northeast and southwest Wollega respectively

Figure 39 this is the preliminary procedure involving ore refinement in north-eastern Wollega (Abee-Dongoro): Packing the ore, washing and the aftermath (the first three from top-left to bottom left) and the bottom right of ore in blocks from south-western Wollega (Aira-Gullliso).

The process afterwards included the selection and further refinement of the ore in one of the surrounding streams in the northeast and packing and carrying it to the smelting sites unlike in the southwest where it was crushed (Figure 39). According to previous smelters, the stream was the first stage where commercial interaction began among miners, tax collectors, smelters, laborers and beggars. The ethnographic re-enactment in 2005 did not include trading in the ore. The following were the procedures involved in the ore procurement and processing in Abee-Dongoro before its final use for smelting.

For the re-enactment (Burka 2006), one of the old miners and smelters led us to the edge of an abandoned pit. Here he showed to the group a mound where the last miners used to accumulate ore before they transported it to the stream (Figure 38 top-left.). Members of the smiths, then, opened up the mound and took the soil into their hands, exchanged from one person to another and it was after they got confirmation from the old miner and the keeper of the site that they filled up the sack.[42]

The required amount of ore was packed in a small sack, which they complained was too heavy for one man to carry it. Hence, two individuals carried it by hanging the sack across a wooden stick. On the way to the stream, they stopped to cut a bark from acacia tree. The acacia bark was spread as a sheet over which they washed, panned, and cleaned the ore from the matrices. It was after the washing that they could be sure whether the ore is enough or not. The washing was not only laborious but needed an experienced person to oversee the process closely, which an ex-slave former smelter did. Otherwise, the proportion of the impurities to the pure ore was so enormous that a great amount of the ore could be washed away by the torrent.[43] After such intensive washing and panning, only 12 kg of pure ore (Figure 39 bottom-left) was obtained from a sack load of matrices.

This procedure contrasts with ore procurement and preparation in southwestern Wollega of the Aira-Gulliso district. Unlike in the northeast where miners were limited to a single iron ore site, south westerners had the advantages of obtaining the ore from four procurement sites to experiment on them alternatively (cf. chapter IV). In addition, the ore in southwest Wollega (Dimma, Wiinsaa Qacoo, Warra Qaalluu, and Caaggoo) is mined in its mass rock form (Figure 39 bottom-right) and is crushed into pieces at the smelting hut. Nevertheless, prior to crushing, they spread the ore in the sun for drying it. An additional refinement included the selection of a strong/rich[44] ore (*kormaa*) from a weak/poor one (*saakure*) by

---

42 Since most of the former pits have collapsed because of age and many are covered with bushes, everyone was warned to follow the trails. In the meantime, the participants showed to me many pits not visible easily due to coffee and other undergrowth. No tree is said to be cut from the canopy forest overhanging the coffee plants and the abandoned mining pits. Ore procurement in southwest Wollega involves very few stages collecting ore, packing and transporting to the smelting site, spreading in the sun for a while and pounding it into pieces during smelting.

43 Note above that the person who examined the ore on the bark spread in the stream has previously participated with his master smelter, and here took part to instruct the second-generation people.

44 There is distinction made between strong (rich) and weak (poor) kinds of ore. The characteristics of the two are not only that the smelters identify them by color but also they feel them while crushing it by hammer. The rich ore (*kormaa* literally means bull) is very strong and the poor one (*saakure* meaning fragile) is easily crushed. It can equally mean that the ore is distinguished as male and female.

an experienced miner-the decision that took place both at the mines and while the smelting was undertaken.

In the case of southwestern Wollega, perhaps due to the quality of the ore or cultural reason, smelters are known to have changed from one site to another. During the reconstruction project (Burka 2008) in 2007, one of the master smelters insisted on bringing ore from a site far from outside of the district of Aira-Gulliso. It was a popular mining site at Dimma in Yubdo [45] (see Cerulli 1933; Hamrla 1966). He tried to justify his argument by relating to a story of one smelter in the past.[46]

His message was that he was disappointed because he could not send a *dibbe*[47] to his lover because Wiinsaa Qacoo's ore failed him (Fn. 46). The song also informs us the significance of iron smelting in the livelihood of the people of the region particularly its role in love, fertility of the society. Despite such accounts, there is clear evidence that local miners extensively dealt with Wiinsaa Kacho ore mining site. This is visible from old trenches and pits all over the field (Figure 38 right). This is one of the mining pits, which previously served people in the environs of Aira for smelting iron.

According to descendants of old smelters in Aira town, where old smelting used to be concentrated, the major source of ore for the site was not anywhere other than Wiinsaa Kacho. It was also possible the ore from Warra Qaalluu and Caaggoo destined to different parts of iron smelting sites in southwest Wollega and beyond. The difference between those in Abayyanii who demonstrated the smelting and those in Aira town must have resulted from the differences and competitions between various lineages (see Burka 2008). Whereas smelters in Aira area claim descent mainly from Laaloo clan, in proper Abayyani village various lineages and sub-lineages as Abayyani, Innagamdi, Innablxii and Malkee (itinerant smelters), whose main occupational identity was smithing, conducted smelting, at least during its later period.

---

45 The ore-mining site is located in Yubdo district, at a locality known as Dimma. His reason, although others were not convinced and argued against it, was that the ore of Winsa Kacho although very close to them (less than one-hour walk in short cut routes) was *saakure*. *Saakuree* is, according to them, an easily crushed ore but full of slag (*balxii*).

46 According to the tradition, the smelter decided to use ore from Wiinsaa Qacoo against advice from his counterparts. He then built his smelting hut, furnace and organized necessary technological variables. After he brought the ore from Kacho he lit his *gaayyaa* (tobacco water pipe) and sat down to supervise his industry. The smelting took place accordingly. After the smelting, he could retrieve a bloom full of *balxii* (slag). Then, he sat down and sang albeit an ironic song, which he addressed to his lover: *gordana Wiinsaa Kacho Qixalamaan saakuree-* Wiinsaa Kacho's ore its inside is *saakure*; *Silaaa dibbee si ergaa qixalamaan isaa uree-* I should have sent you *dibbee* piercing its inside.

47 a pot drum with skin as a diaphragm used by girls during traditional songs Providing this argument the master smelter convinced the rest of the group and me to pay for the trip to Dimma, a distance of over 4-5 hours across the rugged hills, in their short cut tracks. I did not accompany them. The team with their four donkeys should wait for two days before they were back to the village. This was in addition to a previous day when I have settled administrative matter between the neighboring district authorities to permit for the mining.

5.3.2. CHARCOAL MAKING

Charcoal is another natural product and an essential technological variable in traditional iron smelting. Iron reduction from the impurities is made possible by building carbon monoxide (CO) to breakdown the oxygen (02) inside the furnace. There are varieties of ways, which traditional smelters and smiths adopted to procure different available hardwood in their locality. Similarly, they have adopted variety of ways to produce charcoal for the purpose of smelting or smithing. Essentially, iron smelters and smiths in Wollega selected similar hardwoods for charcoal making (Figure 40). The reason might have been that smelters and smiths in both regions claim common origin and that the distribution of the hardwood species in Wollega is much similar. The names for the hardwoods are equally the same in both areas but differ in the mechanism of making charcoal.

Nevertheless, there are other minor differences. For instance, in the northeast smelters used the *baddeesaa* tree for smithing claiming that it 'dries' up rather than 'gluing' the particles together. Therefore, according to them charcoal made from *baddeessa* tree is used to forge the bloom into tools at the smithing stage. They also prefer the hardwoods they call *hambalta* or *biirroo* (*Entada abyssinica*) of the Mimosoideae family and *bosoqa* or *hudduu fardaa*[48] (*Trena orientalis)* of Ulmaceae family for smelting. In the southwest, the choice is somewhat different. According to them, *baddeessa* (*Syzgium guineense*) of the Myrtaceae and *hambaltaa* or *biiroo* trees are used for smelting and charcoal from *hudduu fardaa/ bosoqa* is used for smithing (Figure 40). The southwestern smiths have also improvised to use other tree species such as eucalyptus for charcoal making. This has probably emanated from necessity to alleviate the problem of scarcity of the hardwood as the law also protects the indigenous trees.

There is also disparity in the means of making charcoal. Generally, there are two ways of making charcoal: in the pit and on open spaces (Burka 2006, 2008; Haaland 1985, 2004b; Haaland *et al* 2000). Both methods are used in Wollega (Figure 41). In the northeast, charcoal is prepared both in the pit and on the open surface. According to the re-enactors, the tradition of the region holds that charcoal for smelting is prepared in the pit as opposed to charcoal for smithing during tool repair. For them, if the activity is concerned with forging of the traditional bloom, charcoal must be prepared in the pit. Contrary to this, charcoal is made in heaps in southwestern Wollega (see Figure 41). There is no information hinting that they have changed this over a period. None of them also could recall a story associating charcoal making with pit.

The procedure of charcoal making included: the selection of the younger branch of the hardwood (old wood or the trunk of the tree is not preferable), cutting them into regular sizes, lining them in the pit (in the northeast) or in heaps (in the southwest) and setting fire

48 The same tree species is known by two different names and for different uses among iron producers in Wollega. Among northeastern smelters, know hudduu fardaa (Trena orientalisis) is appropriate for smelting not smithing. On the other hand, in the southwest it is known by another name called bosoqa. The other difference on this particular species is the fact that in southwest it is used in smithing setting to enable the forging of the bloom into tools not in reducing the ore into bloom.

Figure 40 the two prominent hardwoods, Baddeesaa (Syzgium guineense) of the Myrtaceae family (left) and Hambalta (Entada abyssinica) of the Mimosoideae family (right)

to them. For the 2005 re- enactment in Abee-Dongoro, charcoal was made from the already fallen trees during the slash and burn farming. The wet Ethiopian summer season was the reason why they decided to make charcoal from already fallen trees. In both areas, charcoal is prepared at procurement site i.e. near the forest and on the edge of streams.

The connection of charcoal mainly with streams is also available from storytellers. When they are asked how smelting was conducted, the first thing they refer to is where charcoal is prepared. They begin '*lagatti galanii, cilaatti gubatanii...*' literally 'residing in the river, burning charcoal for oneself...' In practice, it refers to the separation of charcoal making sites away from the village where the inhabitants are located. The significance of *lagatti* is that it distinguishes between what is 'wild' and what is 'domestic'. On the other hand, the reference to a river is very essential as evidenced from ethnographic information during re-enactment and reconstruction of traditional smelting. Their explanation is the practical need for water to quench the charcoal after it was half-burnt. The other reason for taking

Figure 41 note that the charcoal maker in Abee-Dongoro is inside the pit beginning to light the fire (left) and the heap of baddeessa tree wood is prepared to be lit for the demonstration in 2007 by Aira-Gulliso smelters (right).

it away from residential areas was attributed to fear of fire accident on thatched houses. Nevertheless, a number of etic interpretations can be drawn.

One common feature between the two sites is that after the wood was half-burnt it is quenched with water. In this case, the intensity of fire was controlled and it avoided total burning down of the wood. In the southwest, they stayed close to the site where they continuously retrieved the burnt parts and proceeded to manipulate them until they were satisfied with the amount they needed for a particular work (cf. chapter III). In the northeast, after the wood was half-burnt, the first measure was to quench it with water. Then leaves were heaped on top of the still burning wood. At the end, they finally covered the pit with soil. They made sure that no smoke would escape. If the smoke escaped, it would mean that the wood totally burns down. Then the pit, which now became a mound, was left aside for three to four days to gradually burn and cool down.

Unlike this, in 2007 when charcoal was prepared for reconstruction in the Aira-Gulliso district, the burning was conducted in one afternoon. During the afternoon, they continued to manipulate it when in the evening they retrieved an amount deemed sufficient for one smelt. However, the estimated amount of charcoal was not enough, as it required them to supplement it from their store. At their store, they had charcoal made from eucalyptus tree. This was stored for smithing purpose. This was an opportunity to observe how people could adopt new means to counter problems. It further shows that despite their assertion that certain tree species were not suitable for smelting or smithing, necessity seems to have forced them to innovate to tackle problems.

5.3.3. TUYÈRES MAKING

Windpipes are other essential technological components of traditional metallurgy. The inlet of air into the reducing environment in the furnace facilitated, which is an important technological variable for smelting, by the windpipes. Tuyères (the windpipes) are commonly made from clay (Haaland 1985, Haaland *et al* 2000; Kense 1985; Sutton 1985; Todd 1985). Ethnographic research both in Ethiopia and outside only focuses on tuyères made of clay (Haaland 2004a, b; Haaland *et al* 2002; Haaland *et al* 2004a, b; Todd 1985, Todd and Charles 1978b). The difference in the description of tuyères lies mainly in their size or the location and direction of its intrusion into the furnace or the depth inside the furnace. Nevertheless, that is more complicated for the Wollega tuyères than its neighboring regions as in the Omo valley (e.g. Haaland 2004a; Haaland *et al* 2000, 2004a; Todd 1985; Todd and Charles 1978a, 1978b).

The first thing to note is in terms of the general setting, namely the number and types of tuyères. In Wollega (including outside the sample for the study region) both smelters and smiths have used and still use the same types of tuyères. My observation of the working tools of a smith among the Amhara north of the Blue Nile in Gojjam shows similar set of tuyères are used as in Wollega. There are three set of different tuyères required to conduct either smelting or smithing: they are tuyères made of clay, which they call *madabii* (also *qunxurroo*), tuyères made of horn (*buudaa, mundoo*) and tuyères made of metal sheet (*kololii*).

However, clay tuyères making involves a variety of technological and cultural aspects than horn and metal ones. These include the selection of proper clay and dumping for 'fermentation' over a period of time, combination of other ingredients to mix with the clay soil for refractory purposes, and observation of some cultural manifestation. As a result, I will provide a more detailed description of the preparation of clay tuyères in Wollega.

The mechanism that tuyères makers in Wollega employed shows the presence of local variation as well as congruency. For instance, according to ethnographic evidence from northeast Wollega, tuyères were made on the same day the clay was extracted. This is different from the southwest where clay needs to be dumped and kept to 'ferment' for about nine days. It is also mixed with chaff and sand gravel. The processing procedure (e.g. mixing with sand) in the southwest is similar to the way a potter processes clay (see Figure 42). In the northeast, the smelter/smith extracted the clay early in the morning and called his working team to commence its making that involved a number of procedures.

The decision of the smelter/smith of the re-enactment project in the northeast was inconsistent with the oral information we collected from old smelters to whom he also listened. However, claiming that through years of experience obtained through trial and error for making tuyères from the same clay for his smithing, he decided to work with it without inclusions and dumping for fermentation' (Figure 43). As a proof, he showed to us a clay tuyère in the forge, which has been used for over a year and endured heat.

Figure 42 a coffee pot maker is mixing the clay with fine sand under her feet. Before this she used the wooden mortar on her left to pound it on the ground.

Figure 43 the madabii tuyères for the re-enactment project in 2005. Note that there were two styles: one of the tuyères was molded around the wooden stick and the other is made free of it to be pierced afterwards.

According to him, the tuyère was made on the same day he extracted the clay and without additives.

A pot maker from southwest Wollega, close to Aira town, can also be taken as another example of improvising outside the rules of local tradition that sanctions certain prohibitions. The potter makes coffee pots on the day she extracted the clay or if she were busy, she does on the next day after she extracted the clay (Figure 42). The rest of the local potters hold that clay must be extracted, stored, dumped in a 'clean' setting from which transformation into a cultural object is gradually made. The 'modern' potter was not worried about a 'polluted' person moving into her working place or about the 'powerful' *Boorana* eyes casting spells on her clay works yet to be consolidated. Her working place is in front of her living house and close to highway where people pass freely. It must be noted, however, that the potter was a converted Christian and was not abide by the Oromo traditional belief systems.

Iron smelters of the present day Aira town area used to extract clay from the same source the potter did but had to process it for longer period before they made tuyères. In any case, the mechanism such as beating or flattening the clay under feet to make more malleable in both iron production sites of Wollega is generally similar. In both areas, a wooden mold is prepared

Figure 44 Madabii tuyères of Hinnika Jare group for reconstruction of traditional smelting in 2007. Note that the cracks. The tuyères were spread on a bed filled with chaff.

in advance, around which the wet clay soil is plastered in a circular mould, or penetrating it with the wooden mould after it was shaped. The wooden mold is needed to create a uniform pipe as much as possible (see Figures 41 and 44). Whereas the former was from the 2005 reenactment in northeast Wollega, the latter was obtained from southwest Wollega in 2007.

In both cases, wooden mold was a requirement. Using the wooden mold, tuyères makers roll the clay around it until the intended shape and size was attained. The other method was piercing the clay mold with the wooden mold to create a pipe. *Madabii/clay* tuyères were made with a wooden stick inside them both in the southwest as well as in the northeast Wollega. One round of tuyères made for the 2007 reconstruction (Hinnikka Jaree site) cracked where they spread them under the shade of banana trees. Participants of the reconstruction offered their respective explanations[49], for the whole of the clay tuyères that cracked.

49 The explanation for the cracking of the clay *madabii* tuyeres in 2oo7 were two: the younger members of the group attributed the incident to lack of proper mixing with the sand soil. They said that too much sand has been included during the smoothening work, which denied it its elasticity. On the contrary, the oldest of all *Obbo* Tafarra Qana'aa attributed the cracking to lack of observing ritual requirements. For him, the tuyeres were made on a market day-the when both 'clean and unclean' people mixed with one another. In addition, he said that the tuyeres were made when the makers were still hot with the alcohol drink they got at the markets. They had also consumed some food as meat, which 'rots' the clay.

Figure 45 Kololii iron tuyères are made from metal sheet after it is flattened and rolled around a mould of its size amongst various moulds in the forge. Note the setting of the forge where kololii tuyère was being made for the 2005 reconstruction.

The other reconstruction site (Laga Hareeraa) a kilometer away did not face the same problem although the source for the clay and the sand was the same. Nevertheless, in the second demonstration site the tuyères gradually dried inside the maker's house on a scaffold close to the roof. According to the master smelter, he denied people's access to the room where he displayed them for fear of pollution, *gaaddidduu*.[50]

There is no standard size, however, for the making of the clay *madabii* tuyères. We observed variation in thickness and refinement based on the experience of the makers. In addition, the size of the furnace determines the length of each tuyère. Generally, the makers and smelters claim that *madabii* tuyères are about 1.50 meters long. This was the case with the tuyères for the southwest of Wollega, which allowed a free space of about 10 centimeters

50 The term *gaaddidduu* literally refers to the shade of some object or person that casts over another object or place or person due to the relative position of the former with angle of the sun. However, in the context of the use of the term in the document it implies the impact of the uncleanness over the object under *gaaddidduu*. This comes from lack of abstaining from different prohibitive activities including food and drinks (see note No.45 above).

from the floor of the furnace for filling the space with charcoal to allow the impurities flow and accumulate there.

The other aspect of the clay tuyère is its representation of a grown man's penis. Hence, the name is *Qunxurro*. The smelter replied to my question for why they had to use pair of *madabii* tuyères for one smelt. He laughed and asked me, 'Don't you like twins'?[51] As I have previously discussed, the size of clay tuyères and the depth of penetration into the furnace is probably connected with the belief of the people about procreation and fertility. There is a tradition (as part of jokes) that big penis is productive and it is so when it goes deep inside woman's genital. Similarly, farmers prefer huge hoe to small one arguing that deep cutting of the soil increases its fertility and agricultural productivity.

The second type of the three set of tuyères in Wollega is an iron tuyère, known elsewhere in the region as *kololii* (Figure 45). The *kololii* tuyere is inserted through the mouth of the bellows to protrude out of the horn, *buudaa* tuyères acting as a bridge between the two tuyères. The need for a metal sheet tuyère is justified by its resistance to the flame of fire that shoots back through the clay pipe. It is permanently installed as far as the bellows were fine. Both smelters and smiths own the tuyères. Relatively, *kololii* tuyères are permanent. I observed this tuyère when the smith/smelter was making for the 2005 re-enactment of iron smelting in northeast Wollega (see Figure 45). For the 2007 demonstration in southwest Wollega, the smiths/smelters transferred the iron and the horn tuyères to the newly made bellows from the old bellows at the forge.

The term *kololii*[52] refers to the smooth flow of substance such as liquid. However, from practical purposes of the smelting/smithing, it may refer to the role of the tool to facilitate an easy flow of the air from the bellows to the furnace. It is a liminal stage between the bellows with cold wind and the furnace, which is extremely hot. It increases the intensity of the air that is blown into the belly of the furnace, thereby minimizing wastage of energy. The fire could come back through the clay tuyères opening and could cause damage to bellows and the blowers if not prevented by *kololii* tuyères. On the other hand, the use of the term *kololii* in association with the flow of liquid or a channel created because its flow can suggest an attribution of the tuyère to the male genital duct, urethra. This ascription becomes clearer during the explanation of the third category of the tuyères made of horn.

The third kind of the tuyères is made from horn, which is known as *buudaa* in the Oromo language. Horn is commonly used for making different drinking cups and decorative objects among the Oromo of Wollega. There are specialized artisans who mainly work on horn and woodcrafts. Socio-culturally, the use of horn objects as cups, is very symbolic. For instance, elders bless couples at marriage carrying a honey brewed drink (mead), *booka,* in horn cups. Since the origin of the cups is associated with the 'wet' animals, it conveys message

51 The cultural reason for making two clay tuyeres was to produce twin blooms, which corresponds to twin babies. In other words, the two long tuyeres (long penis) were to impregnate the woman furnace to give birth to two children at once, an explanation to go with the tradition of the society favoring more children.

52 In love songs by the boys, (e.g. *kololii koo*, meaning my *kololii*) it refers to smooth body, behavior, and the fragility of a woman. It can also refer to a moment of sexual climax.

of wish of fertility for the couples. The use of horn is not limited to its domestic utilitarian purpose as drinking cups of home brewed alcohol but was used by the smiths/smelters as a windpipe.

The term *buuda* (or sometimes *mundoo*) connotes, however, more than its common use (see Figure 46). First, *buudaa* refers to the raw horn, which is yet to be processed. It specifically refers to the sharp side (distal) of the horn. Other than this, it connotes an uncircumcised man's genitals/*mundoo*. Furthermore, people use the term to refer to an uncultivated 'tongue' or rude behavior. The process of cutting its tip end to allow the *kololii* tuyère to come out through it can be regarded as an act of circumcision or an act of taming wild behavior.

The *Buudaa* or *mundoo* tuyères making is concomitant with the iron and bellows sewing. Contrary to the clay tuyères which are made over many days (5-10 days) before the day of the smelting and many days after the furnace was erected or constructed, the making of horn and iron tuyères can take place just a day before the smelting. The making of horn tuyères is not that complicated. The smelter selected the horns from his reserves and estimated the required size. He then cut out the sharp side with an axe (Figure 46).

During the reenactment in northeast Wollega (Burka 2006), horn tuyères were made on the day the master smelter scheduled to finish work on the bellows. The bellows were sewn with the wooden slat along the mouth of the bellows. The wooden slates are for closing and opening the diaphragm for inhaling and exhaling the air. Simultaneously, the master smelter arranged and firmly tied the horn and iron tuyères with a leather strip with the nozzle of the bellows. Both horn and iron tuyères were inserted through the mouth with the distal protruding through the bottom holes (Figure 47).

The number of iron and horn tuyères needed for smelting in both sites corresponded to the two bag bellows operated by one person. Each of the bellows had one opening at the base through which horn and iron tuyères protruded. Since four bellows are required for

Figure 46 Note the tuyere maker was measuring the part of the horn tuyere to be cut off (left) and he was also sewing the mouth of the bag bellows (right) with a wooden slate used to operate for opening and closing the diaphragm. Also note that the horn tuyere is already inserted into the bag to be seen through the bottom aperture

Figure 47 note the setting of the tuyères and bellows together. The other particular observation is the recent change on the use of the bellows. The bag bellow (on the right side) is made from the inside of the car tire.

one smelting and since each bellow had one opening at the bottom, the number of horn and iron tuyères needed for one smelt is eight altogether. With the two clay tuyères, the total number of tuyères required for one smelting was therefore ten. For the smithing however, two bellows, two horn tuyères, two iron tuyères and one clay tuyère make the number of tuyères required 5. This is half of the number of tuyères needed for smelting (see Figure 47 for the arrangements). While the previous two tuyères were the same for smelting and smithing, the clay tuyère is much different particularly with regard to its size and the care given for its making and maintenance.

As presented above, the Oromo smelters/ smiths used bag bellows. No record is found among different parts of the Oromo region where smiths used bellows other than bags. We recognize its deep-rooted cultural phenomenon in their transformation or modification of the material not in the design of the apparatus. Thus, it is possible that they have not changed the type of bellows they have been using for long period. Despite the recent change from skin bellows to car tire tubes, the smelters/ smiths still operated on bag bellows (Figures 47 and 48).This is unlike smelters and smiths of the Omo valley where other kinds of bellows were the technological variables in both smelting and smithing activities (e.g. Haaland *et al* 2004a, 2004b).

Figure 48 one of the smelters/smiths of the village where the reconstruction was conducted prepared the bag bellows for the purpose. Note that he was trampling under his feet on a flat timber after the material was dumped for 3-4 days.

In the past, tanners known as *Duugduu* used to make bellows for smelters/ smiths. Smelters/ smiths claim that the tanners used to make bellows for them because they depended on them for the scraping metal tools for working on hides and skins. It was then a mutual relationship based on mutual interest. However, in time, smelters/smiths learned to prepare their bellows by themselves. It is probably due to such overtaking of their job by the smiths and lack of market for their traditionally made leather mats, containers etc that there were no tanners in the villages or in other surrounding villages where metal crafting is widely practiced in both study sites in Wollega. During our reconstruction of smelting in southwest Wollega (Burka 2008), bellows were made inside the smelting house by one of the smiths/smelters (Figure 48).

As I have mentioned earlier, both smelters and smiths use bag bellows for their work. The Oromo term for bellows is *Qalqala buufaa*, literally meaning blowing sack. In the smelting setting lying on both sides of the blower, the bellows symbolize his 'sacks', the testicles. If one says *qalqala kee*, literally your sack-he is referring to the testicles. During the smelting, the workers challenged one another by referring to parts of male reproductive organs and by using other sexually laden gestures. In referring to the Oromo bag bellows, I did not, however, exclude the possibility of their using other versions of bellows similar to their neighbors in the Omo valley. It is likely that the bellows passed through evolutionary developments.

Many scholars of African archaeometallurgy (e.g. Chirikure 2005; Cline 1937; Haaland 1985; Sutton 1985; Todd 1985) discuss that bag bellows are the higher level of attainment by African bloomery smelters and smiths. In this case, other bellows such as pot bellows or wooden drum bellows are attributed to earlier phases in the development of the wind diaphragms. Similarly, I am not arguing that Oromo smelters and smiths remained stagnant in the technology of making bellows. One ethnographic example regarding the improvising of bellows making is observed in the recent changes from making bellows from goatskin to the use of inner tube of tires.

Smiths in Wollega are now widely adopting the tire tube for bellows (see Figure 47 left). They found that the tire tubes are durable if well protected from the reach of fire. It is also cheaper[53] than goatskin bellows. Unlike goatskin bellows, which they had to carry morning and evening to the residence and to the working place, tube of car tires bellows can remain permanently in the forging hut. According to them, dogs or other wild animals or pests could not cause damage to the tire tube bellows.

5.3.4. FURNACES OF WOLLEGA: FURNACES WITH AND WITHOUT SUPERSTRUCTURE (*GUMBII* AND *BOOLLA BUUFAA*)

The furnace is the major technological variable of traditional iron smelting. It is the first to be made only preceded by the house in which it is erected (cf. *chapter III*). It is relatively a permanent property of the smelter involving many days of treatment and passage through various critical stages. The critical aspects for furnaces include both physical and non-physical ones. The rest of the activities surrounding the smelting were not isolated from the nature of the furnace. Almost all components of iron smelting are liable to the nature of the furnace. The labor organization for the final day is also part of the *chaîne opératoire* of metallurgy.

The construction of a furnace begins with forfeits/ sacrifices for the appeasement of the diverse spirits. For the northeast re-enactment in 2005, the master smelter slaughtered a chicken and when the smelt failed, some old smelters attributed it to the fact that since birds are moved by wind it means that the ore was 'dispersed' rather than 'collected'. Nevertheless, for the smelting demonstrations in southwest Wollega in 2007, goats were prepared according to the accepted tradition of the smelters. Sheep are far from recommended to appease malevolent spirits such as by the malevolent spirits.

Young male or young female goats were not also offered for the appeasement of the sprits during smelting. They are called *duudaa*, which literally means 'deaf', and means 'virgin'. This is common practice in day-to-day conversations when referring to unmarried girls, *durbaa duudaa*. It also connotes innocence and purity. Therefore, only mother goats are

53 According to smelters/smiths in southwest Wollega, the cost for one tube from car is between 25-30 Birr while one bellows from goat skin costs over 80 Birr.

Figure 49 an old granary left where it was erected in a house before it was demolished. It can either be transferred to the new house or broken and mixed with clay to make another new one. Note also the different sections and marks

offered. They recommend *re'ee gara saaqqatte*, literally *goat that got her stomach operated* implying mother goat. For them, sacrifice of a mother goat ensures the productivity or fertility of the furnace.

The other divergence between the two sites was in the building technology of the furnace apparatus. Smelters of the two sites employed drastically different technologies in furnace construction. This is unlike the rest of technological variables such as tuyères and bellows where only minor differences are discerned. The furnace in northeast is without a superstructure. Nevertheless, the furnace in southwest Wollega has a superstructure. In Abee-Dongoro district, the furnace is prepared under the ground in sloping position at a depth of over one meter and a half. On the contrary, Aira-Gulliso smelters construct the furnace from the surface upwards like a granary (see Figure 49). However, furnace building begins from about half a meter in the ground.

The other variability regarding furnace in Wollega is the name by which it is known. In Abee-Dongoro the smelting furnace is called *boolla buufaa*, which literally means blowing pit (*boolla* means pit and *buufaa* is blowing or bellows). On the other hand, in southwest Wollega of Aira-Gulliso district, the furnace is called *gumbii* (see Figure 49). *Gumbii* is a traditional granary made of clay plaster. It is woman's domain. *Gumbii* is sectioned into different parts with anthropomorphic names. It is prepared before the main harvest in early winter. The main base is called, *haadha* (mother) and the sections are *ilmaan* (children). The nodules on each of the parts are called *mucha* (breast or tit). Other body names include *hudduu* ('bottom'), *garaa* (abdomen), *afaan* (mouth), *hidhii* (lip) etc.

Thus, since the furnaces in the two smelting districts are different, despite the claim by the southwestern Oromo of their descent from northeast Wollega and that they carry similar clan names to the Oromo of the northeast, it is clear that both passed through different technological trajectory. However, in both areas smelting and smithing are conducted by people who belong to separate occupational identities-smelters and smiths. A change of identity from one to the other is not common. Nevertheless, in both southwest and northeast Wollega, there is enough evidence indicating the combination of smelting and smithing activity particularly by the smiths. Unlike in the southwest, here is no evidence in the northeast of former smelters becoming smiths.

The common tradition is that smelters did not know how to forge the bloom into tools and did not have a forging site and tools. In other words, the two activities were situated in different localities (cf. chapter II). On the other hand, in the southwest, there is an instance of the family members of a well known smelter becoming smiths and today one of his sons is credited with forging highly regarded bush knives. Except for this separate incident, old smelters categorically deny that their predecessors had any skill of forging tools and ascribe this role to the occupational identity called *tumtuu*. Despite their insistence in dissociating themselves from the *tumtuu* group, some members of the mainstream regard both smelters and smiths as *tumtuu* while other members of the mainstream refer to smelting as a practice by people 'like us'. There is clear distinction between the mainstream and the artisans as exemplified by 'them' and us'.

5.3.4A. PIT FURNACE (*BOOLLA BUUFAA*) OF ABEE-DONGORO, NORTHEAST WOLLEGA

The study of technological variation in iron smelting was conducted by investigating the different production apparatuses. One of the production apparatuses in iron smelting technology is furnace. Scholars divide furnaces into two broad categories. These include slag tapping and non-slag tapping types (e.g. Sutton 1985; Tylecote 1975; cf. chapter

III). Further divisions, however, include furnaces with superstructure and without them. Furnaces are also divided in terms of the mechanism to charge air into them. As a result, furnaces are categorized into natural draught and forced draught (see Cline 1937; Schmidt 1997; Schmidt and Childs 1985; Schmidt and Avery 1987). As I have previously discussed technological variation is also viewed from the extent/depth of penetration of tuyères into the furnace (Schmidt 1997; Schmidt and Avery 1978; Schmidt and Childs 1985; Todd and Charles 1978b).

Previous ethnoarchaeological and ethnographic enquiries in different parts of Ethiopia mention the following categories of furnaces. Accordingly, all furnaces had superstructure but varied in terms of sizes ranging between 80 cm to over 2 meters tall (Cline 1937; Haaland 2004a; Haaland *et al* 2000, 2004a, 2004b; Huntingford 1955; Lewis 1965; Todd 1985; Todd and Charles 1978a, 1978b). In addition, the clear observation of furnace attributes is in the divisions into natural and forced draught furnaces in Ethiopia. The significance of the furnace of northeast Wollega is mainly that no one has recorded its type elsewhere in Ethiopia and the Horn (Burka 2006). Only a remark on the pre-Roman Meroitic furnace has similar attributes (Tylecote 1975). This raises an interesting question about the origins. However, since the purpose of this chapter as well as the whole thesis is not about origin, I concentrate below on the description of the ethnoarchaeometallurgical evidence (but cf. *chapter VII*).

Old smelters who instructed the young volunteers to reenact the traditional smelting in 2005 held that the furnace was prepared close to streams under a high ceiling hut meant to avoid fire accident. They claimed that the hut was a high roofed one to prevent the flame of the fire from burning the hut as it had a thatched roof. For this similar purpose, they said, the smelting site was located away from human habitation areas.

The verbal instruction during my interviews of old smelters (Figure 50) and the descriptions of the technological setting included the following. Smelting hut, *godoo buufaa* was a high ceiling open wall hut. Underground furnace, *boollaa buufaa* was prepared inside the hut in a sloping position approximately over 150 cm in depth. Four bag bellows, *qalqala buufaa*, were made from huge male goatskin, *tumaalessa*.

The other components of smelting, tuyères were made from three materials and known with different names including clay (*madabii*), horn (*buudaa*), and iron (*kololii*)). Accordingly, clay tuyères were made by molding around a wooden mold after mixing and kneading the clay with sand, chaff, and slag fragments. The second category, iron tuyères were made from a metal sheet by smiths and the horn (*buudaa*) tuyères were prepared from this material.

The two surviving members of old smelters from the senior Malkee lineage (Figure 50 left) did not cooperate to demonstrate the former smelting activity. They attributed this to their old age; both brothers were over ninety years old. The rest of the eyewitnesses of old smelting (Figure 50 right) were not consistent with the operational sequences of iron smelting as they were the group of smiths who had chance to observe smelting activity and the young volunteer smiths/smelters rejected their statements. The re-enactment, therefore, depended on the information we gathered from the two old smelters and the knowledge, which used to be passed to the young from their respective families.

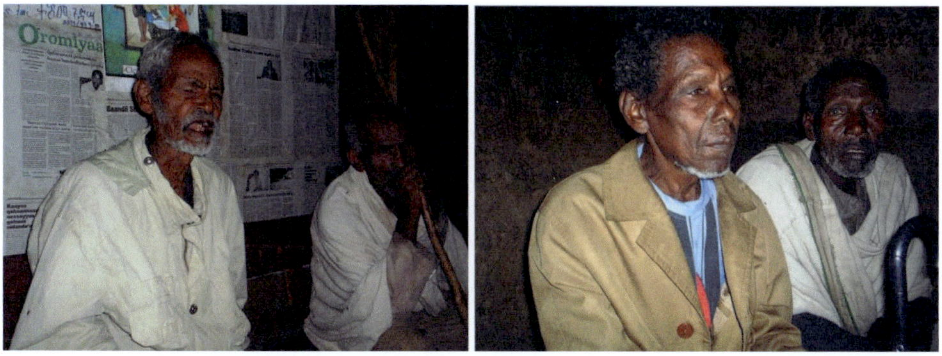

Figure 50 Informants[54] on the past mining, smelting and smithing activities in Abee-Dongoroo district

One of the volunteers was a smith by the name of Darbush Kumsaa from the Yamalagii lineage. He is a grandson of chief of the smiths, *balambaras* Libdii. His father also conducted smelting activity, which I previously argued that was probably in the post-WWII when the Malkee smelters began to drop the activity. While the preparation was under way since June 2005, I conducted survey of sites associated with iron production. In the following month, the furnace and other components of iron smelting were prepared. *Obbo* Darbush took the position of master smelter/smith and organized his team. The Yamalagii smiths dominantly inhabit the Masoo village where he decided to conduct the re-enactment of iron smelting (cf. *Chapter II*). He agreed to conduct the re-enactment under an open hut where he forges the tools. An ex-slave who served Darbush's father and participated in the past smelting was included to assist the young master (Figure 51 right).

The pit-furnace (Figure 51) oriented in a west-east direction and conforming to the slope of the place. The smelters were to sit west of its mouth to pump downwards towards the east. According to the smelter/smith, the slope of the land dictated this. It is appropriate, according to them, to sit on a higher ground for blowers to pump downwards. The master smelter/smith of the demonstration did not comply with many aspects of the instructions he received from old smelters and his assistant in the preparation of the furnace. He was supposed to clean the inside wall from soil in the first place. This was followed by plastering the inside wall with wet clay. Instead, he simply burnt its inside with logs of wood and cleaned the ashes. He said this equally serves with the clay wall building (Burka 2006). Similarly, he did not note the significance of narrowing the mouth of the furnace with clay and building a protective wall in front of the blowers from the flame of the fire. He improvised his project.[55]

---

54 Informants on the left belong to the Malkee lineage and the family who controlled the mining sites, who mined and distributed and conducted smelting. On the right were from men of Yamalagii lineage who obtained blooms from the smelters and forged them into tools. After the former smelters stopped the activities, the latter have erected their own smelting hut. The old man on the further left from the Malkee clan is now deceased.

55 The role I played here was to remind him the words of the old smelters about the requirements in furnace building as well as tuyères making from my field notes. However, he told us all that the old informants are not dependable. Since the assistant, *Obbo* Nagash Feyisa was an ex-slave;

Figure 51 Furnace was built in a circular position under the ground. It has a slopping position extending underground for about 1.50m. The young smith with his tutor decided to build it close to his forge.

It was in this circumstance that the first smelting was conducted in July 2005. The smelting was a failure. It was half cooked ore because no enough heat was produced in the furnace. The smelter/smith attributed the failure to the rainy season as well as broken clay tuyères. Thus, he asked to redo it for the second time, which took place in October 2005. In late September of the same year, a better preparation was made with clay tuyères and the same furnace.

5.3.4B. FURNACE OF AIRA-GULLISO (ABAYYANI, LAALOO, MALKEE): FURNACE WITH SUPERSTRUCTURE

The survey in Aira-Gulliso district in southwestern Wollega revealed that iron smelting has been conducted in several places mainly by the smelters of the Laaloo clan who inhabit

his words did not have weight to convince *Obbo* Darbush. In addition, I wanted to passively participate and record the significance of the improvising by the young members of the group. In the mean time, I could make out that the difference might have originated from the lineage difference. Yamalagii lineage was under the control of the Malkee lineage in the past. It appears that Darbush wanted to show that he could proceed on smelting without the technological provisions of the legitimate owners of the knowledge.

around Aira town (Burka 2008), a former missionary station. Existing evidence from informants attests that the furnace that was used for smelting was of a similar type all over Aira-Gulliso district. This was due mainly to the fact that furnace and clay tuyères makers in this region might have moved from site to site. In addition, the skill was passed from father to son in the form of apprenticeship.

During the 2007 smelting reconstruction in Abayyani locality about seven km south of Aira town, two individuals (from Innawangii and Innagamidi lineages) who had previously conducted furnace and tuyère making were invited to prepare the apparatuses. Accordingly, one (*Obbo* Waratii Fayyisaa) from Yubdo district and the other from Aira town (*Obbo* oliiqaa Lamuu) helped furnaces building at Laga Harera and Hinnikaa Jaree sites respectively. The purpose of conducting two smelting reconstruction simultaneously at different localities was part of an enquiry to investigate the presence of minor variations in iron smelting. In addition, it was dictated by competition between heads of the two minor lineages (see Burka 2008).

My communication and agreement that started in 2004 was with old smelters/smiths of the Innawangii lineage at Abayyani locality. They then agreed to take responsibility to organize the reconstruction of smelting. As things went on, they sub-contracted with the two people who were well acquainted with smelting particularly furnace building and tuyères making. The smelters with whom I have directly dealt with and maintained the same relationship with were those who at some time in the past combined both smithing and smelting but who belong to the *tumtuu* occupational identity. In addition, two sons of former smelter from Malkee lineage who settled among the Innagamdi smiths also agreed to participate.

On the other hand, the two persons who were sub-contracted belong to the smelters' occupational identity of the Laaloo lineage in the mainstream. They claimed they know nothing about smithing as well as all members of their family. They asserted this by stating that it was not 'their nature' but 'theirs' i.e. the *tumtuu*. This suggests that there were known furnace builders and tuyères makers who were always available to individuals who could afford to finance iron smelting.

The whole period leading to a smelting was scheduled to take place in about 38 days (Burka 2008). The period included the construction of a smelting house, clay extraction, ore procurement etc, up to and including smelting. Furnace building was preceded by the construction of a smelting house. Unlike in Abee-Dongoro where smelting took place under a simple open walled hut, the smelting in Aira-Gulliso was conducted in a medium sized family house with a high ceiling dome shaped roof (Figure 52 right). The house is a circular structure of five meters in diameter and 19.5 m tall.

The temporary pillar in the middle of the floor separates the house into two equal parts. After the construction of the roof at Hinnika Jaree was completed, the pillar was cut above the wall and was supported by a horizontally laid wooden beam. However, the group at Laga Hareeraa left the pillar intact and constructed the furnace close to it. In the traditional thatched roof houses of Wollega, the central pillar is where the family fireplace is built, and is particularly of male domain. This hearth is a common meeting ground of all family members although according to tradition women were associated with another fireplace at rear part inside the house.

Figure 52 the smelting house in southwest Wollega is similar to any duelling house. All the inside settings are made in accordance with a countryside thatched roof farmers' houses.

In this traditional countryside house, the main hearth is dug in bowl shape and beaten clay is used to build it. Three stone supports called *goomjii* were erected. One of the three stone supporters, which is perpendicular to the front door is associated with the woman's fertility spirit, *Maaramii*, hence *gomjii Maaramii*. Women perform various fertility rites in front of this stone surrounding the hearth. For example, in the ceremony of rite of passage after delivery, which is performed on the fifth day of the delivery, friends from both wife and husband sides gather and prepare a traditional food made of barley flour and drink, eat and sing ritual songs and challenge one another. The stone is anointed with butter and other ritual materials such as evergreen grasses, etc.

The purpose of this explanation is not to discuss about the significance of the hearth. It is to show the value of the place *vis-à-vis* other parts of the house and its symbolic significance. The smelting furnace is located exactly where the fireplace should it was for living purpose. As I have already discussed, the construction of the furnace was made within the agreed time framework of 38 days as set by the organizers of the smelting. In addition, as I have previously mentioned in the statement of the problem and the background to the dissertation, the Aira-Gulliso project (2007) finds its root in the previous similar research project I initiated in northeastern Wollega.[56] The purpose was partly to increase the vivid picture of regional history of metallurgy.

The building of furnace in southwestern Wollega involved a number of procedures in addition to house construction as a prerequisite. The furnace here is called *gumbii* (granary). In fact, it has the shape and traits of a granary. The questions, which remained to be answered, include, for instance, why both are called by the same name? Does it mean that iron production originally rooted from the purpose of production of agricultural

56 The project was funded by SIDA-SAREC/AAU thematic fund made available to the University Staff on competitive bases. The fact that my proposal was accepted and passed peer reviews was an encouragement for the topic and for a region underrepresented in the many fields of research in the country. Be it as it may, the reconstruction project of 2007 has increased our picture of metallurgical history of Wollega but above all revealed the presence of a diversified wealth of information on the subject.

Figure 53 Pit in the smelting house where clay is dumped for fermentation and imixed with the straw

implements? Alternatively, does it mean that iron smelting was originally an occupational activity of the farmers? On the other hand, does it have to do with the role of women in iron smelting in the past?

The furnace was constructed from the same clay for *gumbii* (granary) by which farmers used to plaster the walls of their houses in the villages around. The name of the clay is called *boolalee*, a leached reddish soil that is not cultivable and is left for grazing along banks of river streams. Even so, for one of the smelting sites we conducted in the villages half kilometers apart from each other, the *boolalee* clay soil was extracted from a coffee farm of an old smelter, who provided guidance in house and furnace construction.[57] They extracted the clay and left it close to the mines for days for 'cooling down' because it was *caariggii* meaning raw and transported it to the smelting site on the donkeys back.

The clay extraction site is about one kilometer away from the smelting reconstruction site. The coordination of the work included: extraction and storing the clay in sacks near the pit, at the same time preparing a mixing pit or a wooden rectangular container in the

57 I therefore paid for his coffee plantation land affected because of mining as well as I did for the owner of the land where the group decided to construct smelting house.

smelting house and transporting to the site. *Teff* chaff was used for mixing with the clay in the 'fermentation'/*bukaahuu*/ pit[58] (see Figure 53).

The furnace was built in a circular shape at the center of the house. A 50 cm string was tied on a peg to form a circle around which the digging took place. The furnace was over 160cm tall when measured from the surface of the ground. The mechanism of its construction is similar to that of the wall of the smelting house. After measurement was made, a narrow circular channel of 50 cm depth is prepared leaving a space for the door. Then, a wooden wall was erected in the channel. The wooden wall was tied together with plant fibers to form the desired shape (see Figure 54 of its reconstruction). Then the channel was refilled with soil. They also plastered it with wet clay to give it more strength. Therefore, the furnace superstructure can be defined as the height of the furnace above the ground.

The description that its building began 50cm deep in the ground should not connote any suggestion that the furnace of Aira-Gulliso was built on a slag pit.[59] On the contrary, the furnace does not have a slag pit said to facilitate the flow of liquid impurities from preventing the proper flow of air by sealing the tip-end of the tuyères. The furnace was first plastered from the outside and left to gradually dry for days. Then the wooden wall is burnt out from the inside leaving vertical channels and subsequently filled with clay plasters (Figure 55).

According to the smelters, the wooden wall is burnt out after they were sure that the clay wall could stand without its support (Figure 54). Previously, I thought that furnace was built with a block of clay in the same way potters make the pots. Two old informants from Aira-Gulliso ascertained that furnace was built without wooden support, which others rejected it attributing this to loss of memory due to old age. The difference among informants suggests that furnace building might have passed through different technological trajectories.

As far as I was concerned, my preemptive conclusion originated from two cases: first, the hitherto information we have regarding furnace building (e.g. Haaland *et al* 2004a; Todd 1985) in southwestern Ethiopia gives such an impression. The second reason was that during my surveys of former smelting sites, I have collected collapsed furnace walls and a macroscopic examination revealed no signature indicating wooden walls were once part of them. However, if subjected to microscopic examination charred channels (Figure 55) left from wooden walls could be identified although successive clay layers hindered visibility for naked eyes. This reconstruction project, therefore, revealed the hitherto unrecorded local tradition of the building mechanism employed by Oromo smelters of southwestern

58 Such a pit seems to complicate archaeological interpretation providing a wrong image that the pit could have been the base of the furnace. The preparation of the furnace totally require about 15 days before it is ready for smelting.

59 As observed from the demonstration by the old smelter, a well known furnace builder and tuyere maker (Fig.4.14), the floor inside the furnace is left without digging but rather gradually smoothened by the pieces of clay left on the floor during the plastering of the furnace wall and there is no specially designed place where the slag flows. However, since the tuyeres were left hanging about 10cm short of the floor where charcoal was filled, the space can serve as a free place where the slag coagulates upon which bloom was formed.

Figure 54 the picture shows the early procedure of furnace construction when an old smelter demonstrates to youngsters with chaff of sorghum (left) and experimental one (right).

Figure 55 the old smelter and furnace builder, Obbo Tafarraa Qana'aa was making the first round plastering from inside after the wooden wall was burnt out. Note the vertical scars left on the furnace wall from the wooden wall. He entered the furnace from the bottom aperture.

Wollega. I am not, however, denying the possibility that some Oromo smelters developed such a technique i.e. furnace with a wooden wall, through trial and error in their lifetime. They might have developed the method in response to recurrent collapse of furnaces built without wooden support or derived it from their vernacular architectural building technique.

The first firing of the inside wall of the furnace was mainly, therefore, to get rid of the wooden wall before the second level of construction was initiated. The wooden ash was then cleaned out from the surface of the furnace wall.[60] Then the wall in the inside was plastered with the same clay soil they had fermented and prepared in the pit. According to the oldest of all smelters, *Obbo* Tafara

Qana'aa (c. 90 years old), the inside part was plastered in three stages. The first was immediately after they avoided the wooden wall. According to them, this was a critical stage, which they called *offolee* when the furnace could pass the test. After they were convinced that the furnace passed the critical stage, they proceed to the second round of plastering with a thinner layer of clay. Then, the third round takes place a day or two days before the smelting. It was smoothening of its inside wall with refinement.

Furnace building passed through various stages implying that it was the core of the iron smelting technological apparatus.[61] One still wonders why the smelters of southwest Wollega developed such a complicated technological apparatus as opposed to the pit-furnace of their close lineages in the northeast. The technological variability observed between closely related lineages of the Oromo of Wollega lends new dimensions of insight on the role of local adaptation and experimentation with technology such as metallurgy in a society.

One of the significant parts of the furnace of southwest Wollega is the front bottom aperture (Figure 56). The front bottom aperture of southwest Wollega furnace serves various purposes. On the one hand, the aperture is through which the furnace maker enters to finish the inside wall (see Figure 56). On the other hand, it is through which the master smelter checks the arrangement of the clay (*madabii*) tuyères during the process of charging. In addition, it serves as technological advantage. It is through the aperture that the master smelters control the rhythm and force of blowers. Finally, the aperture is through which the smelters drag out the bloom and the discard including broken or half-burnt tuyères, slag, and charcoal.

After the final plastering was complete, the furnace was left for some time to consolidate gradually. The experts regularly examine for possible cracks and seal them. The night

60 According to an old smelter (Figure 55) inside the furnace the purpose of cleaning wooden fragments from the furnace inside wall once it was burnt was to prevent the furnace from burning that gradually shortened its lifetime through the medium of the leftover of the wooden walls.

61 As we see from the photo, (Figure 55) the man is sitting in the furnace cautiously plastering the wall from inside. This is immediately after the first firing to remove the wooden wall. It can be seen from the picture that the wooden wall has left clear traces, which are filled and sealed with clay. The man inside (*obbo* Tafarra) is assisted by another knowledgeable son of a friend of *Obbo* Tafarra who inspects the developments and provides the clay from outside through the aperture (Fig.56).

Figure 56 the assistant of the master was providing clay mixed with chaff to the master inside the furnace. Note the size of the furnace at the bottom to allow a man to easily moving inside it. At the bottom, the furnace is 50cm radius but narrows upwards.

before the smelting day is the final test for the furnace. As a result, huge fire was lit inside the furnace. The fire was kept burning until the morning, and the ash was cleared before charcoal was filled. Some fire is left inside the furnace to ignite the charcoal, however. According to old smelters, the pre-firing of the inside wall takes place only for a new furnace. Similar procedures would follow when an old furnace was repaired. For them, the furnace was a permanent technological apparatus. Nevertheless,depending on cracks or roughness of its inside through use, occasional plastering for smoothening its inside was made. The complete and ready for use furnace can be seen with anthropomorphic features of over 160 cm tall from the surface of the ground and from its base where building began it was about 210 cm. The furnace of southwest Wollega is a female one (Figure 57).

The description of 'her' body (*qaama ishee*) parts corresponds to parts of woman's body. The furnace has a head-like projection depicting an afro-like shape (*goofaree*) (Figure 57). On both sides of the head (*mataa*), it has two ear-like (*gurra*) projections. The upper wide aperture is known as mouth (*afaan*) with the lower edge of the opening representing the bottom lip (*hidhii*). At the far back inside the mouth, a shape was fashioned to imitate the glottis (*qoonqoo*) and epiglottis (*huuba qonqoo*) (Figure 63). The shape that formed the 'glottis and epiglottis' of the furnace, came from two openings (*ijaa*) on the back of the

Figure 57 Front side of southwest Wollega furnace. The smelters have nick-named it Goobanee. Note its anthropomorphic body parts and see the description (Figure 56).

shoulder for insertion of the clay tuyères. On the upper part of the front side of the chest are two parallel protruding nodules (*harma/mucha*) to depict full breasts of a pregnant woman. Below the breasts is an outwardly pushed belly (*garaa*). Since the general posture of the furnace is in a back leaning stance, the outward pushed belly creates an impression of a pregnant (*ulfaa*) woman (Figure 57) and its sketch (Figure 58). An aperture is left below the 'belly' at the bottom on the 'front' side. The aperture's symbolic representation of woman's genital, *faaliqaa or koshee* becomes clearer only during the process of the smelting.

As I have previously mentioned, on the back shoulder of the furnace immediately below the head and parallel to the mouth circular openings are left for the insertion of the clay tuyères. The tuyères formed a right-angled triangle with their tips pointing to the front aperture, *koshee*. Such a top-down position of penetration for the tuyères is not known in Ethiopia and the Horn (see Haaland 1985, 2004a; Haaland *et al* 2000, 2004b; Haberland 1959; Huntingford 1955; Lewis 1965; Todd 1985). Such a procedure is recorded in western Africa (Killick 1999, 2004). Thus, unlike, for instance, the Tsara furnace recorded by Haaland *et al* (2004a, b) and Todd (1985) for the Dime smelters in the Omo valley of southwestern Ethiopia, the arrangement of tuyères for the Oromo smelters of Wollega is

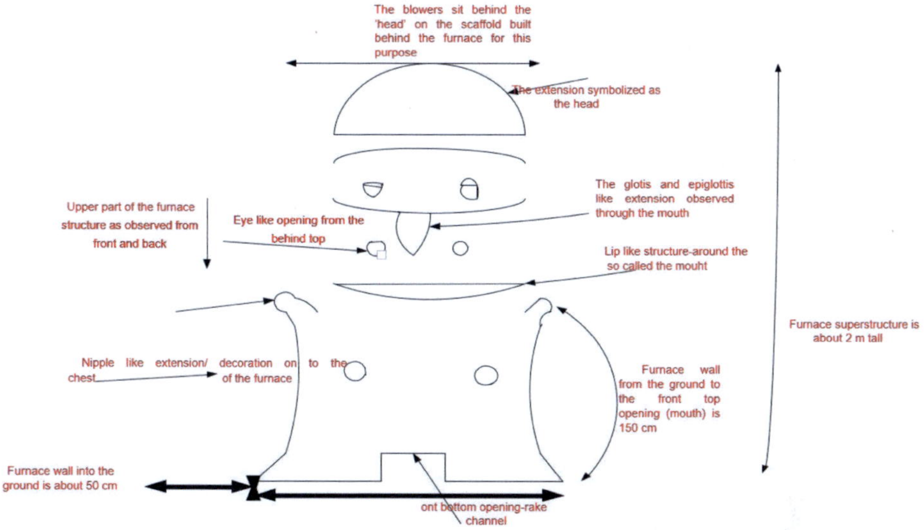

Figure 58 Note the sketch of the above furnace is according to the names given to different parts of the furnace.

from top down. As very well recorded for the Tsara and Dime smelters, 5-6 openings are left surrounding the base of the furnace through which tuyères are inserted. Although Todd (1985) claims that the tuyères of her ethnographic record were said to have penetrated deep to allow preheating, they were not over 30 cm long.

The southwestern as well as the northeastern Wollega tuyères (clay) are over one meter long fully planted inside the furnace. The tuyères for the 2007 reconstruction were 150 cm long allowing a free space of about 10 cm between the tip of the tuyères and the floor of the furnace. In addition, the fact that the insertion mechanism is from top through the belly of the furnace down makes it technologically viable in terms of 'preheating' and separation of the slag, which is different from the neighboring regions.

The expression I made above that Oromo furnaces were operated from 160 cm above the ground raises the question of the bellows' operating mechanism. The smelters of southwestern Wollega also devised a compatible solution for the problem (Figure 59). From the position of the openings, it becomes apparent that blowing operation takes place from the 'backside' of the furnace. Here then comes the significance of the head-like extension above the mouth and the openings on the shoulder. Its practical significance in addition to its symbolic importance was to protect the bellows and blowers from the radiance of the fire that come out through the mouth. As a result, bed was built with proportional height to the shoulder opening behind the furnace. Therefore, the two blowers sit side by side to work on four bellows connected to a pair of five tuyères (Figure 59).

Figure 59 two bellows workers have seated themselves on the bed built at the back of the furnace parallel to the two shoulder openings.

In general, before the smelting time, all the above technological and other socio-cultural requirements were fulfilled. Once the smelters were assured that the necessary elements of production have been set in place, they would assign people who took different responsibilities with the master smelter. The main activity during smelting was to work on the bellows. Other minor activities included pounding the ore into small pieces (southwest Wollega), slaughtering a sacrificial animal, mixing clay to seal the openings, selecting charcoal, engaging in chorus of love songs to boost the morale of blowers, etc.

5.3.5. THE SMELTING

The smelting procedure in both sites of Wollega was generally similar. In both areas, the requirements discussed above were more or less fulfilled before the smelting day. The major difference lies mainly in the furnace apparatus. On the other hand, the type and number of bellows, the type, the size and the number of tuyères and the mechanism of charging charcoal and ore into the furnace allow us to draw the conclusion that the smelters in both regions shared many common practices. However, the span of time one smelting took in both regions was still different. In addition, according to informants, smelting in northeast

Wollega was conducted only once in a day.[62] On the contrary, in southwestern Wollega, smelting was conducted twice in one day. In the northeast, after smelting was over, the furnace was left to cool down gradually throughout the night if smelting was conducted in the daylight or throughout the daytime, if smelting was conducted during previous night. Nevertheless, in the southwest, old smelters held that both rounds of smelting were conducted between morning and evening.

During the ethnographic reenactment in 2005 and reconstruction in 2007 demonstrations, I have observed the fact that the previous accounts are generally certain. In the northeast, smelting was started at six in the morning and the master declared it over after seven hours of sweating labor. Then, it was left aside until the next morning. The furnace was cleaned from the remnants of tuyères,charcoal, and ash. The master and his assistants collected a blackish mass (Figure 60 from the second round smelting) from the bottom of the pit-furnace.

However, the young master smelter/smith who took the role of coordinating the smelting demonstration was not able to transform the smelting product into a viable object. The smith is known to be the most skilled among the smiths in the region who inherited his skill from his well known father. The special bush knife Darbush Kumsa produces from scrap iron is the most sought by the farmers of the region. Thus, what happened to the bloom/slag of his smelt? Why did he fail to forge a tool out of it? The re-enactment of Abee-Dongoro was previously described as unsuccessful (Burka 2006). The reason for such a conclusion was found in the smelter's/smith's failure to forge a tool out of the coalesced material not in the lack of product from smelting. As far as producing wrought iron/bloom was concerned, the smelters have accomplished their tasks. It was due to lack of experience to deal with the unsorted material, for persons who used to work on scrap iron throughout their experiences. Thus, *Obbo* Darbush's reasons for difficulty to forge them became clearer during the reconstruction work in southwest Wollega in 2007. However, the smith has offered his explanation for failure to forge a tool out of the bloom.

Firstly, he blamed the failure on his lack of previous experience to work on indigenous bloom, which mainly explains why he failed to transform the smelt into usable metal. According to experiences from southwest Wollega, such a stage needs not only a skilful smith but also a person who is highly patient to manipulate the material by heating, beating (very lightly at the beginning) and preparing it for forging. Secondly, the forge (of Darbush) did not have enough charcoal to heat the bloom/slag to the level of malleability. In a way, given the improvising measures the smelter/smith took from the beginning to the end, the re-enactment activity opened a number of vistas of information about the technology and other aspects of indigenous iron production in northeast Wollega.

Contrary to that of the Abee-Dongoro site, smelting in Aira-Gulliso was well organized and thought out. It was a reconstruction, by actual smelters and participants in the

62 One day in Ethiopian context implies 12 hours. Accordingly, in northeast Wollega smelting which according to informants took between 7-10 hours took place either during day or throughout the night. In southwest Wollega, smelting took about 5 hours and according to old smelters since the second round immediately follows the previous smelting, two smelts were usually conducted between sunrise and sunset of the same day.

Figure 60 the smelters of northeast Wollega were checking on the unsorted bloom retrieved in the morning. They were not confident of their success and could not make a tool out of it. Those who smelted (the smiths) never worked on a bloom, which was stopped before their time

past. Persons who in the past conducted their own smelting activities fully handled the procedures. Many of the assistants had also participated in one or other areas of smelting. Consequently, the knowledge of smelting was still a vivid memory among many elder inhabitants of the village including the women. The project for 2007 organized 36 people (ranging from young smiths to old smelters) to participate at various stages of operational sequences of the smelting. The project also provided an opportunity for the young and the old to exchange the knowledge, which they dropped over the last 40-50 years.

Iron smelting in Aira-Gulliso was a swift activity. It began between seven and eight in the morning. Between seven and eight o'clock, some activities including arrangements and breakfasting were made. One of the arrangements was to insert the two clay tuyères appropriately overlooking the front aperture. Then the back openings were sealed with wet clay to keep the tuyères firm in place and to prevent smoke or flame coming through them. The master also examined the proper position of the bellows *vis-à-vis* the clay tuyères. Then charcoal was added first through the bottom front aperture, *faaliqa*. The space between the floor of the furnace and the tips of the hanging tuyères was carefully filled with charcoal up on which the tuyères lay. When the master and the assistant were assured that everything was in order, they began to seal the bottom aperture.

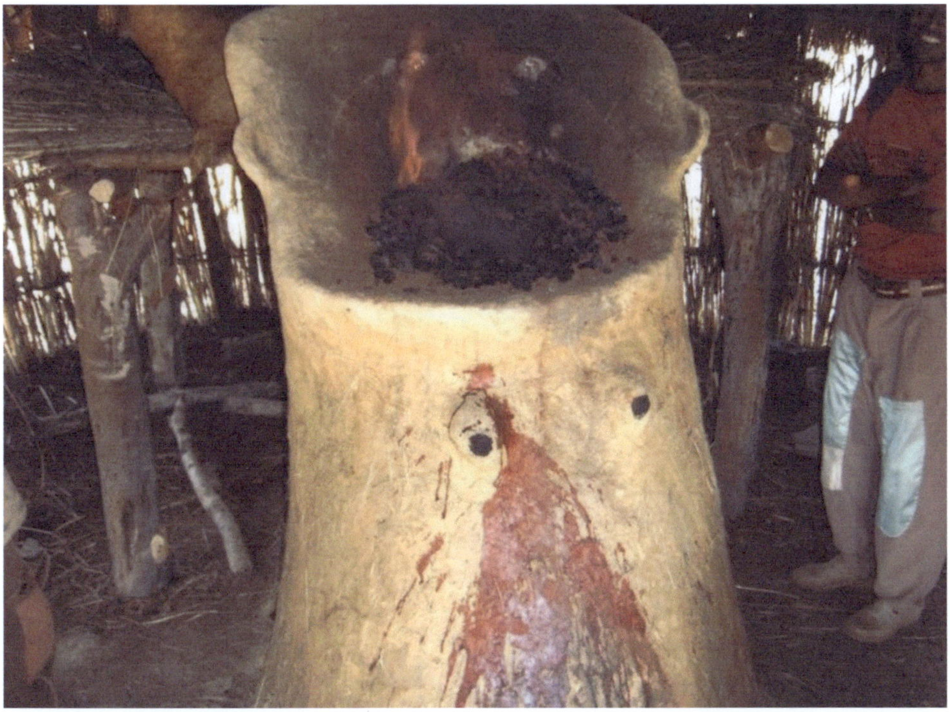

Figure 61 this was a setting when the furnace top was lit with fire and the front side of the furnace was anointed with goat blood.

The aperture was sealed with vertical alignment of broken tuyères and clay. Then, the furnace builder opened a circular hole with his forefinger. Charcoal was then added through the mouth with layers of ore being spread. Charcoal and ore were added in layers. Crushed pieces of iron ore were spread on the surface of the charcoal and the procedure was continuous until the charcoal was filled to the 'mouth' of the furnace. A small fire, which was left inside the furnace, began to create smoke. Additional fire was added through the clay tuyères and at eight o'clock, blowing was started.

The two master smelters made regular checking on the position of the bellows, the blowers, and the alignment of the two types of tuyères attached to the bellows with the clay tuyères in the furnace. In the meantime, smoke began to escape through the mouth from the charcoal. Then one of the master smelters lit fire on a handful of grass and took it to the 'mouth' of the furnace. Flame caught the entire region of the mouth as if it had been charged with a gas (Figure 61).

Meanwhile, the goat's throat was cut and blood was brought in with a container and then smeared between the breasts of the furnace and down the belly. In the meantime, prayers were made by the old smelters and spectators responded affirmatively echoing the words. Everybody should say 'yes' (*eeyyeen*) and 'be it' (*haata'uu*). No negative or

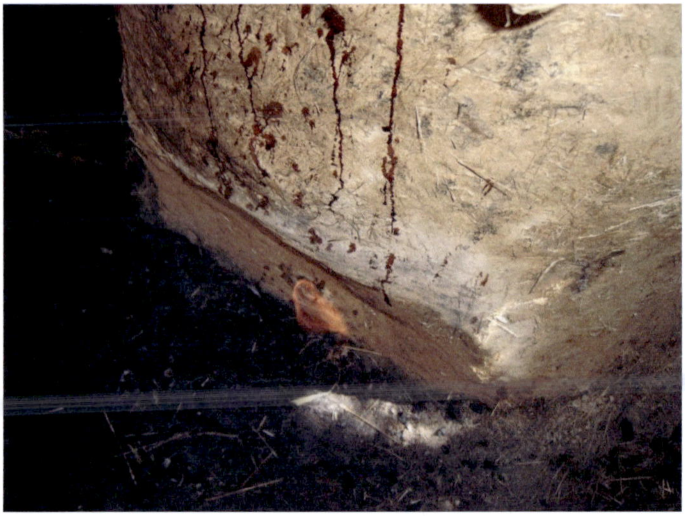

Figure 62 Note that the flame of fire is protruding through the peephole, faaliqaa. When there was no interruption in the rising and falling of the hands, the flame's 'tongue' sticks out through the faaliqa.

Figure 63 the 'glottis and epiglottis'[63] part of the furnace. This picture is taken through the 'mouth' of the furnace

63 The apparatus is called *Qonqoo* and *huubaqonqoo* (glottis and epiglottis). The smelter who decorates the epiglottis answered to my query of the need to beautify it with a mocking smile. "Does it not look like 'her thing'?" He said the 'epiglottis' represents the clitoris of the vagina.

pessimistic statements were allowed at least at the initial stage. It is forbidden to suggest that the smelting would be unsuccessful or the tuyères would break, etc. Then, the masters concentrated on supervising whether the bellows were properly working.[64] There was a mechanism to check on each of the blowers' seriousness or even on each of the hands. Rhythmically, smooth transition between the two hands operating on two bellows was needed. Proper hand coordination between the co-blowers was also important and examined by the uninterrupted sticking of the flame through the *faaliqaa* (Figure 62). According to master smelters, balance (particularly the strength of their puffing) between the two blowers was essential.

The apparatus (Figure 62) that helped to leverage the blowing mechanism is called *faaliqa*- it is a colloquial term connoting a peephole left at the previous aperture. They also call it *koshee* referring to woman's genital organ. The peephole is the one which was left after the front aperture through which the master entered to plaster the inside wall and the first layer of charcoal was fed (Figure 62). The strength of the blowers is also measured from the length of the flame 'tongue' sticking out of the peephole. In the meantime, the goat's meat was served. Primarily, the master took pieces of roasted meat and threw it to different directions inside the house. It was an offer for the spirits of the land and the ancestors. Blowers and other participants also made the work animated by singing love songs such as I quoted in connection with iron ore and charcoal in an earlier part of this chapter.

One other important technological parameter, used by the smelters during the work in progress, is the glottis and epiglottis (Figure 63). In addition to its symbolic role depicting part of the human body, its technological significance is that the smelters used it to estimate time for charging charcoal and ore. They frequently peeped through the flame and asked *huubn qonqoshee mul'atee*? Did her epiglottis show up? If the answer is yes, they begin to fill first the charcoal and then the ore and finally charcoal. They use phrases such as 'full to her lip' (*hamma hidhiitti*) and 'make her gasp' (*ittansiisi*). Simultaneously, two people waiting for their turn relieve blowers who were through. Therefore, the epiglottis (*huuba qonqoo*) is not just a decoration. It is rather a technological variable in the traditional iron smelting technology.

In these iron smelting reconstruction and re-enactment practices, the aforementioned procedures continued until they claimed they heard the 'snoring' sound echoing from the tuyères in the furnace. The change in the sound of the tuyères comes due to its shortening after the distal part of it melts into the fire. At this point, the master smelters announced that the smelting was over. This was done after 13 rounds of recharging with charcoal and ore and blowers had exchanged for the same number of times. In their experience, they remember to have continued up to 20-21 rounds. According to them, this depended on the amount of bloom they sought to produce. It was logical in that smelting was for commercial purpose. They stated, however, that the same sound in the tuyères signaled to stop smelting.

64 Then, the masters intensively observed the supervision of the work. Eight blowers were identified for their skill and vigor. They were paid for one day.

Figure 64 the procedure of bloom retrieval through the rake channel

The claim that they have gone up to 20-21 rounds is not convincing in view of the ethnographic observation that revealed that the tuyères had melted half way during the five hours of 13 rounds of charging in five hours time span. Nevertheless, as the span of time between each charge gradually narrowed, as the furnace got hot, and since the old smelters claimed that smelting took place in 4-5 hours, it is possible that 20 or 21 rounds could be achieved by the same size of tuyères and time span. This is particularly possible when the furnace is already dry or an old one.

The other important technological aspect of the smelting was that during the 5 hours long smelting, blowing has never been interrupted. The practical reason offered was to prevent the flame from shooting back through the pipes and cause damage to the bellows and blowing bed, which is made of grass mattress, and even worse burn down the smelting house. The intensity of blowing only reduces during the period of recharging. How was it, therefore, possible to quit when the master announced the smelting was over? The mechanism to control the backfire of the flames through the tuyères openings was to seal with wet clay mould waiting on the top of the furnace. This clay roll was put into the opening simultaneously when one blower stops one hand while the other was still working.

The process of 'delivery' followed this. The delivery mechanism included reopening the bottom aperture that was previously sealed after charcoal was fed through it. When the seal of *faaliqa* was broken, a glowing red charcoal oozed out of the opening (Figure 64). While one person pulled out the mass of glowing charcoal with a stick, the rest of the people smeared water onto it. Part of the fully unburned charcoal was selected for future use. When the charcoal mass was cleared and the smoke from the quench subsided, it became possible to see coalesced material hanging along the tuyères still attached partly to the back wall and the suspending tuyères.

The master smelter declared that the bloom was a twin. Then the hot mass was knocked off with a long stick from the hanging tuyères to be withdrawn through the *faaliiqaa*. In contrast to the northeast Wollega example, in the southwest, bloom was withdrawn while it was still red hot (see Figure 65). When it finally came out, a plant climber was brought to drag it out of the house through the '*faalliqa*' door, a phenomena symbolizing child delivery. It is interesting to note the marks such procedure left on the ground all the way between the furnace and the place in front of the smelting house (Figure 65) could be an important archaeological signature. Outside of the house, some of the slag was knocked off and the rest is left to be sorted out at the smith. The ethnographic example provided a perspective on how smelters used to manage the spaces following the smelting.

After this, the smelters/smiths collected the unsorted bloom fragments to the smithing hut of the coordinator. The smithing forge was repaired for this purpose (Figure 66 left). Another quantity of charcoal was prepared for smithing the bloom. It was, however, not so easy to forge a tool from the bloom (see Burka 2006, 2088). One of the problems was that it changed into molten liquid after it was reheated (Figure 66 left). As a result, it was proposed that a cautious means of handling was required. In this way, both the master

Figure 65 the still hot bloom mass after it was dragged through the front door of the furnace. Note the climber used to drag the bloom. It symbolizes mother-child separation mechanism after delivery.

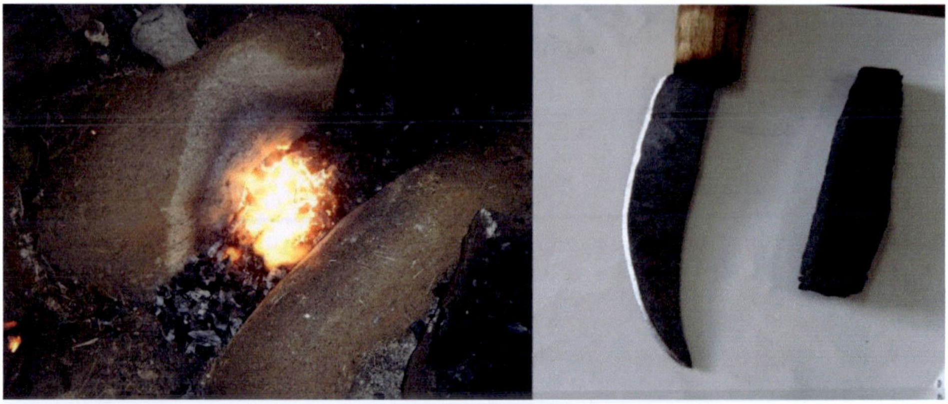

Figure 66 the smith's forge prepared for smithing the bloom (left) lit up with glowing bloom and a knife and an ingot made from the ethnographic reconstruction at Hinnikka Jaree smelting site, Aira-Gulliso (2007).

smith/smelter (*Obbo* Kumarra) and his advisor (*Obbo* Tafarraa) agreed to leave it cool down until the next morning.[65]

In sum, the ethnoarchaeological presentation in this chapter has provided us with many insights into the procedure followed in indigenous iron production in different parts of Wollega, western Ethiopia before they stopped the practice about half a century ago. Iron production clearly supported the local economy in different ways. Description by 19th century literary sources that indicate that iron was traded outside immediate region is evidence of its surplus production above local consumption.

Within Wollega, however, iron was destined for not only secular use to satisfy the socio-economic and military needs of the people but also trickled to traditional religious sites for ritual purposes (Burka 2009; cf. *chapter VI*). On the other hand, the origin and role of the artisans are still among the issues for further investigation. Some of these intricate problems are among those I will deal with in the following sections of the dissertation.

Some of the questions I will ponder on include, for instance, how far the myths of origin of the smiths portray evidence on why in the first place the innovation of metallurgy needed. The following chapter would redress the socio-cultural aspects of iron metallurgy left out in previous chapters that dealt with the technological apparatus.

65 The smelters discussed and agreed to leave it to cool down and to be deal with next morning. In the next morning the master smelter's cowrker, who is much credited with his refined skill in forging various knives took the seat and forged a number of bars. I have taken one of the bars with me. They asked me to forge into objects of my liking. I told them that they could forge a knife. I asked them how many objects could be made from the blooms. They said some four knives. They blamed on the tuyeres that were broken inside the furnace. The second smelting was successful until the end of the smelting. However, since the master, just to impress me with his success tried to mix with old bloom fragments they had with them from the past smelting, I advised him not to forge any tool.

CHAPTER SIX

6.0. THE SOCIALITY OF IRON: THE MATERIALITY OF THE MYTH OF ORIGIN OF CRAFT SPECIALIZATION, IRON WORKIG AND MARGINALIZATION OF THE *TUMTUU* OF WOLLEGA

The only way in which relations of domination can be set up, maintained or restored, is through strategies which, being expressly oriented towards the establishment of relations of personal dependence, must be disguised and transfigured lest they destroy themselves by revealing their true nature

Pierre Bourdieu (1977:19 in Freeman 2001:302)

6.1. INTRODUCING MODELS ON THE ORIGIN OF MARGINALIZED ARTISANS IN ETHIOPIA

This chapter broadly deals with the materiality of socio-cultural aspects of iron metallurgy. It explores how the materiality of metallurgy has intertwined and embodied with the perception of the society about the trade and its skill holders. In this regard, first, I introduce the literature on the models to describe the status of the various craft workers in the Ethiopian region and the Horn. Then, I attempt to situate the Oromo traditions in light of the various models and evaluate the relevance or even the peculiarity of the data from study area.

Thirdly, I discuss two competing traditions on the origin of iron smelting in Wollega from smelters and smiths perspective. The presence of contrasting traditions reveals the presence of internal struggle between the groups on the ownership of knowledge and by implication a contest over history. Fourthly, I present the ritual role of metals among the Oromo of Wollega. In this last part, I concentrate on the changing role of metals (between secular and spiritual) as their context changed. I have obtained the data for discussion on the ritual role of metals from two ritual sites of the *Qaallu* (Burka 2009) in northeast Wollega (districts of Horro Jimma and Haro Limmu). The discussion generally encompasses Wollega and its environs.

I have started this chapter by quotation from Pierre Bourdieu (1977 in Freeman 2001) to convey the challenges in making out the structural relationships in the Oromo society of Wollega. I believe that it enlightens us with the mechanisms to maintain relations through transfiguration of certain aspects of identities such as occupational differences. The significance of such a statement is found in the need arising from maintaining and ensuring continuous relationship and interdependence among the various socio-economic and cultural sectors of the society. To maintain this interdependence, there has to prevail a social organization that keeps different social groupings in their separate places.

Put differently, the social organization in place makes sure that separate identities and differences were consistently maintained by disguising the identity of each group through social or spiritual sanctions. One can posit that such mechanisms are of panhuman characteristics (see Haaland 1985, 2004b; Haaland *et al* 2002; Haaland *et al* 2004a). The questions are thus: Why are separate identities of work needed? Alternatively, what are reasons for keeping occupational identities apart? How are they kept apart? It is significant to find whether universal explanations were there for the separations. Haaland *et al* (2004a:80) have to put such a phenomenon in that

> *No matter the scale of the political systems, the association of crafts with specific ranked identities was an important principle of social organization in all of them, although the occupational tasks associated with specific identities as well as their relative rank, varied with scale and complexity of the political centers.*

Drawing insights from the analyses of marginalized craft workers in different parts of Ethiopia by various scholars (e.g. Fleming 1976; Freeman 2001; Haaland *et al* 2004a, 2004b; Haaland and Haaland 2007; Levine 1974; Pankhurst 1999, 2001; Todd 1977, 1978) I will try to contextualize the socio-cultural aspects of metallurgy in Wollega. I concentrate particularly on major attributes of the materiality of marginalization expressed through occupational status. In addition, I focus on other attributes as ranking and endogamy that impose certain types of restrictions (either social or supernatural) on each of the rankings. In this case, I will try to identify whether these attributes helped to confine various identities within their separate categories, for instance, settlement. My focus here is on the occupational/social identity known as *tumtuu.* The reason for my focus on the *tumtuu* arises from their association specifically with iron metallurgy and the fact that members of the group contained other identities as well.

Thus, since the discussions by various scholars on craft specialists is generic and that in Wollega the term *tumtuu* itself is generic (cf. *chapter* II) the models created for explaining the origin and life of craft specialists are generally applicable in the context of my study area. The major models discussed for the emergence and existence of marginalized handicraft groups in the mainstream society include caste model, evolutionary or submerged model, internal differentiation or class model and symbolic models (see Haaland *et al* 2004a; Levine 1974; Pankhurst 1999, 2001; Todd 19771978).

6.1.1. THE CASTE MODEL

Advocators of this model explain the idea using different adjectives such as low caste or outcaste in trying to draw similarity or difference between marginalized groups in Ethiopia and caste groups in Asia, India (see Pankhurst 1999, 2001). However, for the proponents of the model caste alone is not enough explanation for marginalization of the artisans in Ethiopia. They assume that there are many other attributes than the Hindu-Indian religious sanctioning of caste (Haaland *et al* 2004a; Lewis 1962, 1970).

On the other hand, Pankhurst (1999, 2001), who traces the use of the model in the Ethiopian context to Nadel's (1954 in Pankhurst 2001) discussion of smiths of East and West Africa disagrees with the use of the model to explain the origin of marginalized craft specialists in Ethiopia. Apparently, therefore, there is no consensus on the applicability of the model in the Ethiopian context. However, others (e.g. Haaland *et al* 2004a; Haaland and Haaland 2007) have come with different ways of applying the model than the religiously sanctioned Indian Hindu caste type.

The alternative means of explanation of marginalization of the group to religious sanction include endogamy, lack of commensality, pollution, ranking and the like (e.g. Haaland *et al* 2004a; Lewis 1962, 1970; Todd 1977, 1978). This is in contrast with the religious ideological basis used by Nadel (1954 in Pankhurst 2001) for categorizing social organizations as castes. However, in spite of the objections, many have already used the model to explain the position of craft specialists in the Ethiopian context (e.g. Cerulli 1922 in Pankhurst 2001; Haaland *et al* 2004a, 2004b; Levine 1974; Shack 1964; Haberland 1963; Todd 1977, 1978).

In addressing the marginalized artisans among the Omotic-speaking societies of southwestern Ethiopia, Haaland *et al* (2004a, 2004b) define occupational identities in the society as caste. According to them, caste is "a social system based on status summation in such a way that having one particular status, for example blacksmith, entails that one also is incumbent of a specific cluster of other statuses, for example, position in ritual, kinship, political structures" (*Ibid:* 2004a: 80). They viewed the existence of separate ranked identities and the prevalence of restrictions to maintain separation between occupational castes, for instance, through classifying them into higher and lower.

According to them, the formation and maintenance of these castes is intertwined with the growth of political power centers whose livelihood depended on the redistributive capability of the political center. In other words, the differentiation of various sectors of the society into ranks is a means of survival of the political center. How is such a model applicable to the marginalization of artisans in different parts of Ethiopia? Is there any universality among different societies on the formation of marginalized minorities? For instance, is the 'Falasha' (see Quirin 1997; Pankhurst 1999, 2001; Tamrat 1972) model suitable for explaining the origin of marginalization of artisans in other parts of the Ethiopia and the Horn?

It is apparent that scholars agree that the 'Falasha' are 'outcastes' (Quirin 1997; Pankhurst 1961, 1992; Tamrat 1972) among the dominant northwestern Ethiopian Amhara society. The marginalization of the group who became the 'Falasha' is attributed to a historical process and transformation that took place in northwestern Ethiopia since the second half of the 13th century (Tamrat 1972). The historical process that led to the creation of the 'Falasha' marginalized community or caste group began with the transfer of power from the Agaw led Zagwe dynasty to the Amhara dominated so-called Solomonic dynasty.

The appearance of the 'Falasha' identity is said to date back to the later part of the 13th century (see Freeman 2001; Quirin 1997; Tamrat 1972). The appearance of 'Falasha' identity has a wealth of political, economical, and religious motives. The Amhara Christian highlanders expanded their rule into northwest Ethiopia legitimized by the ideology of the Ethiopian Orthodox Church.

The culmination of the campaign to evangelize the region mainly came during the period of Emperor Yishaq (see Tamrat 1972). The campaign coincided with Bet-Ajir's, (one of the Ayhud

Agaw leaders) refusal to pay tribute. Taddesse Tamrat (1972:200-201) summarizes, the historical context under which the 'Falasha' identity was forged, in the following manner:

> *Yishaq fully realized the essentially religious nature of the Falasha question. For him, the only solution to this chronic problem lay in bringing an end to the religious problem...he passed a decree from which the Falasha...have derived their name: 'He who is baptized in the Christian religion, may inherit the land of his father; otherwise let him be a Falasi...[since]then, the House of Israel have been called Falashoch [exiles].*

According to Taddese Tamrat, therefore, the refusal of the Agaws to be part of the Amhara Christian kingdom is rooted in their realization that the system will deny them their traditional religion. Apparently, however, they were overwhelmed by the Amhara power legitimated by the church. While the Amhara encroached on the occupied territories building churches and occupying fertile areas, the 'Falasha' were more and more pushed to the margins and predominantly became craft workers (Quirin 1997; Freeman 2001).Their status changed with the changing socio-economic system in the region.

The 'Falasha', as masons, benefited from the foundation of Gondar as permanent capital in the 17th century. However, the decline of central power in the subsequent century resulted in the dwindling of construction activity. This forced the 'Falasha' to revert to their old works as blacksmiths, weavers, and potters (Freeman 2001; Quirin 1997; Pankhurst 1961; Shack 1964). During this later time, they were tasked with performing undertakings including amputating the limbs of convicted criminals, treating infectious wounds, and skin diseases (see Freeman 2001), which are regarded by others as polluting. In this complex historical context, the 'Falasha' evolved from low-ranking groups created by Amhara domination to caste-like position with various stereotypes attached to them as *buda*, possessors of the evil eye (Pankhurst 2001; Quirin 1997).

Although various means of classification in a society are also used to define a caste group (e.g. Haaland 2004a; Haaland *et al* 2004a), some argue that the model only holds for Asian societies (Pankhurst 2001). According to Alula Pankhurst (1999, 2001), the ideological, genetic and structural difference between Asian and African societies implies that the model will not work in Ethiopian contexts. He further on maintains (Pankhurst 1999) that the main objection is ideological that focuses on the lack of religious sanctions on the part of Ethiopian societies. In addition, the difficulty to ascribe caste to entire society and the fact that the relationship between marginalized and mainstream society in Ethiopia is on 'patron-client' basis than on the exploited and exploiter makes the explanation using caste model far from plausible (Pankhurst 2001).

Even so, for Dave Todd (1977, 1978), the system that operates in the Dime society, for example, can unequivocally be called caste. Similarly, as I have pointed out earlier, Haaland *et al* (2004a:80) defined caste from the perspective of the existence of ranked occupational identities whereby ones category under one identity entails other statuses to be maintained in separate existence through strict restrictions imposed on them.

In this case, separate existence sanctioned by restrictions based on such as pollution or high and low created various identities. With regard to the Omotic Dime society, Todd (1977) identifies the presence of hierarchy in top and bottom layer divisions divinely approved and beliefs about pollution prohibited access between divisions (Haaland 2004a; Haaland *et al* 2004a). The categorization into high and low is evident in activities such as sacrifices whereby the marginalized craft workers were not allowed to carry out this ceremony on mountain tops (Todd 1975, 19777, 1978).

From these perspectives, it is possible to categorize occupational identities in Wollega as caste groups. As I have discussed under *chapter II*, Oromo society in general is structured in ranked hierarchies putting one on top of the other, *gararraa fi gajjallaa*. However, this is in spite of the emphasis by scholars on the society's 'egalitarian' characteristics (See Hassen 1990; Kassam and Megerssa 1989; Legesse 1972; Megerssa 1994 for the egalitarian issues and Burka 2006 for arguing against this). There is the social organization in Oromo that divides people in to top, *gararraa* and low, *gajjallaa* in terms of habitation, sacrifice and social gatherings. The question is if we categorize the *tumtuu* identity as caste, we need to look into a criterion that meets the characteristics of the group. Alternatively, did the *tumtuu* identity put the group members above or below others?

In Wollega, ideological factor cannot be used to explain the marginalization of the *tumtuu*. Therefore, other explanations must be sought for why members of the Oromo mainstream ostracize the *tumtuu* group (Burka 2006, 2009). As explained in Pankhurst (2001), the *tumtuu* do not necessarily form a single clan let alone a society. However, they are still different as their existence is not generally on patron-client relationship. Even then, despite the presence of rank, it is less elaborate and pervasive and that marginalization is moderate enough to allow close relationships between the people (Burka 2009).

I argue that the marginalization of craft people among the Oromo of Wollega lacks roots in the world view or philosophy of the people and is therefore superficial and might have been a recent introduction (see Burka 2009). The separation between these identities is made relevant through restrictions on the commensality and marriage on the grounds of purity. However, there were instances where marriages could be forged breaking the ranked identities. It is true; however, that such incident was 'healed' through a number of ritual interdictions. In the following section, I, as well, will present another model in order to draw possible scenario for the ranks among the Oromo of Wollega.

6.1.2. THE REMNANTS (THE 'ABORIGINAL NEGROID') OR EVOLUTIONARY MODEL

Southwestern Ethiopia became open to travelers and explorers in the 19th century, who were impressed by the presence of marginalized hunters and gatherers 'isolated' from the mainstream society. The question regarding the identity of these people prompted these travelers to forward various models for explanation (Pankhurst 2001). One of these models was the 'sub-merged model', which argues that these groups must have belonged to 'remnants' of people who were engulfed by waves of conquerors (see Freeman 2001; Levine 1974; Pankhurst 2001; Todd 1978). However, this model can be challenged on various grounds. If they were of different ethnic origin, is there a way of discerning them from the mainstream? Do they acquire cultural or physical features that distinguish them from their 'host' communities? Do they, for example, speak different languages from their host communities? Do all marginalized people have any tradition claiming of common origin? Though Shack (1964) attempted to distinguish the Fuga from the mainstream Gurage society based on skin complexion and some argots, he found no linguistic basis to attribute to them a separate ethno-linguistic identity.

In spite of the aforementioned challenges, a number of works on the culture-history of Ethiopian societies continued to adopt the 'remnants' model as elaborated by the 19th century travelers. According to these scholars (though not all are uniform), marginalized handicraft groups in different Ethiopian societies, belonged to pre-Nilotic and pre-Cushitic 'aboriginal Negroid' peoples who remained unassimilated into the conquering dominant groups (Cerulli 1922, 1957; Haberland 1964 in Todd 1978; Jensen 1959; Lewis 1962, 1970; Trimingham 1952). A similar argument is also found for the Oromo artisans in Wollega.

According to some authors (e.g. Gidada 2001; Ta'a 1980; Waqweyya 1988) a number of clan names in different parts of Wollega, which are today associated in the main with craft working, were identified as 'remnants' of pre-Oromo societies that were 'submerged' by the Oromo conquerors in the sixteenth century. On the contrary, however, there is no evidence including physical features that indicate their separate origin or distinguish the craft workers from the mainstream Oromo society in Wollega.

Although many of these scholars depend on the evolutionary theory, some (for example Cerulli 1922 and Haberland 1962 and 1964 in Pankhurst 2001), differ from the rest of proponents on the question of whether the marginalized communities and the mainstream society have a common ethnic origin. Despite the fact that some physical differences are discernible (Shack 1964), this cannot hold tight since the physical features of the Ethiopian societies are very diverse.

Shack (1964, 1966) pointed to the presence of physical difference between the marginalized Fuga handicraft groups and the mainstream Gurage society. According to him, the Fuga resemble the 'Bantu Negro' (see Todd 1978). Other scholars also found similar explanation for the Manjo marginalized groups in Kafa (Huntingford 1955; Orent 1969 in Todd 1978). One can reject, however, this argument by noting the fact that not all Gurage or Kafficho have similar physical features.

In addition, the model lacks universal applicability to serve as workable model in different parts of the Ethiopian region and the Horn. For instance, how one can explain the *tumtuu* community of Wollega in light of the 'Aboriginal Negroid' model? The clear departure from this model is that most of members of the *tumtuu* of Wollega have a fair skin complexion. Even then, there is no uniform skin complexion between members of the *tumtuu* of Wollega or members of the mainstream Oromo.

Moreover, none of the members of this group (forgers, weavers, potters) speaks language or argot distinct from the rest of population. On the other hand, the artisans claim to have common myth of origin with the rest of the Oromo population. From myth of origin of the *tumtuu*, for instance, it is learned that they claim equal place of seniority, *hangafumma*, in the Oromo society. In other words, they attribute their origin to one of the first Oromo twins created by Waaqaa.

Generally, the model is criticized for lack of universality including archaeological, linguistic, and physical evidence and due to its rigidity and simplicity (see Todd 1978; Lewis 1962; Levine 1974; Pankhurst 1999, 2001). Thus, another alternative model is proposed to explain the ethno-genesis of the marginalized groups in Ethiopia and the Horn. This model is known as 'ecological niche' and 'internal social differentiation' (e.g. Levine 1974; Todd 1978).

6.1.3. INTERNAL SOCIAL DIFFERENTIATION: 'HOLISTIC' AND 'MUTUALISTIC' SPECIALIZATION

This model attempts to project backwards on the materiality of social differentiation in the Ethiopian region and the Horn. Two differing views are entertained to address the issue of how social identities were formed in the Ethiopian region and the Horn. Donald Levine (1974) in his *Greater Ethiopia* espoused the first view. According to him, distinct groups that settled in certain ecological niches developed certain specialized crafts and then dispersed or were incorporated into host groups to sell their trade. Levine thus argues in favor of a flexible approach to explain the presence of marginalized craft groups in different parts of Ethiopia and the Horn (see Fleming 1976; Todd 1978).

As opposed to views by other scholars, which he called 'rigid approach', Levine inclined to draw on a variety of explanatory parameters including genetic, regional, linguistic, religious, ecological or social structure for classifying the various Ethiopian ethnic groups from north to south (see Levine 1974: 33-39). For him, this is a flexible approach as opposed to relying on one type of model. On his flexible' grounds, Levine (1974:166-1969) developed two models for explaining the emergence of 'outcaste' groups in the social evolution of Ethiopia. The two models he called are 'Holistic' and 'Mutualistic' specializations.

Donald Levine's 'Holistic' specialization schema (*Ibid.* 1974: 166-167) involves the creation of various modes of subsistence in diverse Ethiopian ecological niches. These modes of subsistence, according to him, are 'hunting and fishing, plow cultivation, pastoral nomadism, semi-nomadic pastoralism, slash-and-burn cultivation, horticulture, and specialized crafts (Todd: 1978: 148). One can admit that culture is not created in the air. Nevertheless, it is too simplistic to argue that human social structure or livelihood is determined by ecological factors. In addition, not all craft specialties need independent ecological niches (Todd 1978:149). Does for instance 'ecological niche' model work for specialized craft workers such as smiths? Is it possible to argue in favor of the idea that the *tumtuu* discovered the knowledge of metallurgy because they settled close to iron ore, *gordanaa*?

The challenge also lies on Levin's (1974:169) 'Mutualistic' specialization when he states that '...a small society no longer relies on intermittent contact with outside groups.... moves to secure [*to sell their skill, which they developed in their respective niches*]...on regular basis by finding a place for supplies within its own home territory...'. One of the critics of this approach is Dave Todd (see his works on the Dime of the Omo valley in 1975, 1977, 1978). According to him (*Ibid.* 1978:149), the question lies on 'if there was little demand for the products of specialists, why did they develop the specialization in the first place?'

Such formulations by Donald Levine might have come forth from his assertion that 'the Ethiopian caste groups do not represent specialization on the basis of internal social differentiation' (*Ibid*: 169). While discussing the rest of the Ethiopian society according to his 'flexible' approaches, he specifically failed to address how the marginalized minorities of craft specialists have emerged among various communities.

Alternatively, Dave Todd (1978: 150) proposes the congruency of 'ecological niche' and 'internal social differentiation' as explanatory model of the existence of marginalized groups in different parts of the country. He argues that both are not mutually exclusive. He summarizes this view as follows:

> *...that distinct groups of people settled in 'ecological niches', developed 'specialized crafts' and then dispersed into 'host groups' in order to market their products...that societies produced relatively stable methods of food production which did not consume all, or even most, of the members time while producing more than a subsistence level of food. As the knowledge of the country improved, resources such as iron ore, animals, honey, clay, and needs became known and people realized their usefulness...*

Todd (1978: 50) further posits that in places where there were sufficient skills and resources as well as sufficient food surplus to support the community, people may have preferred to live mainly or totally on what they could secure in exchange for their own products. Therefore, internal stratification in societies takes place or is determined by economic factors rather than cultural ones (see Pankhurst 2001). Todd also adds that internal classification presupposes surplus production and a highly developed agricultural system.

It is very important to evaluate the story of the *tumtuu* in Wollega in light of the aforementioned factors discussed that account for differentiation within a society. According to them, though simplistic, all occupational groups were in the distant past from a single family who agreed to share different works according to their creativity. This labor division gradually evolved into a further level of identity whereby the descendants of each group were ascribed to it by birth. As a result, the sons and daughters would automatically take the skill of their fathers and mothers. However, the discussion on how social differentiation/stratification evolved among the Ethiopian societies is not sufficient to explicate why some sections of occupational groups were pushed to the margin of the mainstream society. The other model called 'symbolic model' seemed to redress the gaps left by the rest of explanatory models.

6.1.4. THE SYMBOLIC MODEL

Mainstream society regards handicraft groups as 'dangerous' and morally outsiders or 'impure' (Burka 2006, 2009; Freeman 2001; Haaland 1985; Haaland 2004a; Haaland and Haaland 2007; Haaland *et al* 2004a, 2004b; Hallpike 1966; Lewis 1965; Quirin 1997; Pankhurst 1961; Pankhurst 1999, 2001; Shack 1964; Todd 1977, 1978). The symbolic explanation on the origin of marginalization of craft specialists emanates mainly from the materials they produced or the techniques they used in the production of various tools. It also emanates from the cultural labels attributed to them due partly to some socio-cultural roles that they play for the mainstream groups (Pankhurst 2001). These include circumcision, initiation, midwifery or healing infectious diseases, etc (see Freeman 2001; Haaland 1985; Hallpike 1966; Quirin 1997).

The religious prejudices related to color and characteristics of materials produced by the craft specialists in Konso of southwest Ethiopia is a case in point as discussed by Hallpike (1968) for the metaphors of their marginalization. According to Hallpike, the mainstream Konso society regarded weavers, iron smelters, and potters as 'dangerous polluters' because of the materials they are identified with. According to him, for the Konso mainstream society cotton is color of bone and symbolizes death; iron smelting is symbolized as cooking earth, and pot making is symbolized as treating earth as food (Haaland *et al* 2004b; Pankhurst 2001).

Nevertheless, this cannot apply outside limited contexts. For instance, cotton or its color does not connote any negative meanings among the Oromo of Wollega. In contrast, white symbolizes hope and promise as evidenced by the tricolor of the cultural dress. However, there are other symbolic meanings attributed to the artisans to distance them from equal commensality with the rest of the population. For instance, the *tumtuu* occupational identity is barred from sexual partnership because they are feared for 'burning' their partners because they are 'too hot'. This might have been due to their association with fire. In addition, they are symbolized as 'barren', 'dry' again probably due to the dry and cold metal they dealt with. The Oromo regard extreme cold symbolizing sterility/death.

Although the different models forwarded by the scholars have been used one way or another to explain reasons for marginalization, they are not comprehensive enough on their own to encompass varieties. As a result, it is possible to explain the nature of the organizational principles by combining these different models including internal social differentiation, caste, symbolic and other models. In the following section, I present some of the various stories prevalent among both the mainstream and the artisans in Wollega to explain how occupational differentiation originated in the society. In this discussion, I will attempt to provide balance to the narrations by different groups, in this case, between the occupational identity called the *tumtuu* and the mainstream agricultural society. Since as I have discussed under chapter II there is not clear separate identity for the smelters, the search for explanation of 'marginal' identities is focused on the *tumtuu* smiths (see Burka 2009).

6.2. ON THE MYTH OF ORIGIN OF THE *TUMTUU* AND THE SUBEQUENT MARGINALIZATION OF THE GROUP MEMBERS AMONG THE OROMO

The various models forwarded above are meant to explain the reasons behind the marginalization of minorities among different societies in the Ethiopian region and the Horn. One of the weaknesses of the models is the lack of focus on the views expressed by the marginalized groups themselves. The models were mainly developed from the perspectives of the host population of the minorities of craft specialists (Pankhurst 1999,

2001). They are narrations of how the mainstream society perceived the craft specialists. The narrations were not how the marginalized minorities explained about the origin of their treatment differently from the rest of the population. Therefore, the following description attempts to redress this imbalance by presenting the perspectives of both the mainstream and the artisan minorities and draw insights for archaeological interpretation.

Stories on the emergence of the *tumtuu* occupational identity among the Oromo of Wollega are diverse. As a result, I selectively present, here, three stories told by the 'marginalized' *tumtuu* group as well as non-artisans or the mainstream Oromo. The first story deals with the supernatural origin of the skill of smithing through one of the 'original two Oromo brothers' from whom ultimately the present different artisan communities descended.

The other story is about an incident that occurred during initiation ceremony of peer groups in the remote past that determined the fate of members of the peer group. Finally, I will present another story relating to an incident during the crossing of a big river after which the law changed the occupational identities of members of the community. The last story is popular among both the *tumtuu* and the mainstream society in southwest study area of Wollega while the other two are mainly popular in the northeast Wollega.

6.2.1. *'HINNA FI FINNA'* (HINNA AND KINSHIP) *'HINNA FI FINIINA'* (HINNA AND RADIANCE OF FIRE)

Members of the *tumtuu* and the heads of traditional Oromo religion, (*Qaallu*) in northeast Wollega tell the first story. This tradition is recorded as part of my research plan to enquire for explanation of clan names with the prefix of *Hinna* (*Inna*).[66] I wanted to know why many of the members of the smiths count lineages with prefix *Hinna*. I also wondered whether there is relationship between the smiths and people who oversee the spiritual activities of the community, namely the *Qaalluu*. Some scholars have listed the clan names with prefix *Hinna* to attribute them to 'remnants' of groups before the 'arrival' of Oromo in Wollega (e.g. Gidada 2001; Ta'a 1980; Waqweyya 1988).

These lineage names are generally associated with artisanship though not all members of the lineages were artisans. In Aira-Gulliso district of Abayyanii locality, I have counted about 54 households out of which less than twenty were described as currently not practicing metalworking. However, the informants from Hinnawangii, Hinnatagii and Hinnabalxii insisted that all members of the 54 households belonged to the *tumtuu* group. It means that the society regards all members of the lineages as artisans despite the fact that some did not practice metalworking since childhood. Thus, *tumtuu* has become a social identity ascribed to individuals by birth.

66 The following clans are inhabiting in different parts of both in the study areas of northeast and southwest of Wollega: (H)*Innanaanii*, (H)*Innawaarii*, (H)*Innashabalali*, (H)*Innataagii*, (H) *Innawangii*, (H)*Innabalxii*, (H)*Innango*, (H)*Innacino*, (H)*Innabaaroo*, (H)*Innasayii*, (H)*Innasharbii*, (H) *innashabalalii*,(H)*Innageedii*, (H)*Innaddeenii*, (H)*innagamdii* etc,.

According to the story, Waaqaa, the Oromo supreme God created two brothers from whom all Oromo have descended. These two twines were Anno and (H)inno. After Waaqaa created them, he endowed each of them with different capabilities and qualities. The focus of this story is on the (H)inno groups and the story is mainly common among lineages with this prefix. The story is told to justify the group's claim to the Oromo root as well as seniority. Accordingly, Anno was given administrative skills to lead his people. Since Anno was the first to touch the land before (H)inno, he and his descendants are regarded as elders (*hangafa*). Those who became the *boorana* class among the Oromo of Wollega are said to be the direct descendants of Anno. On the other hand, (H)inno the second to touch the land (*Quxusu*) was given other qualities, which Anno did not possess. Waaqaa endowed (H)inno with two powers namely prophecy (*Milli'uu Waaqaa*) and radiance or sparkle (*Qaanqee Waaqaa*).

Gradually, the two powers were separated into those who would foretell and those who would make ritual objects to those who foretell. Originally, according to this myth, the two powers were given to and practiced by one person namely (H)inno. In the end, however, the powers were divided into two among the descendants of (H)inno. Those who foretell (*Milli'uu Waaqaa*) became Qaalluu[67] and those who got the radiant power became the *tumtuu.* Still, there are areas where one person contained and practiced both powers.

The story further relates that it was (H)inno who first forged the first *Kallacha* from the metal of Waaqaa (metal from sky).Through *Kallacha* and the spiritual guidance of the descendants of (H) inno, Anno and his offspring, the *boorana* were able to establish rule within their society and maintain it peacefully. In this arrangement (H)inno used to make the scepter, *bokkuu*, the symbol of power for Anno. According to this story, since Anno was powerless without the spiritual guidance from (H) inno, the two remained side-by-side on equal grounds. Nevertheless, it seems that Anno's position is upper in the rank than that of (H)inno or descendants due to the right of the first born among the Oromo society. Similarly, (H) inno possessed spiritual power and skill to transform things that puts him above Anno.

Thus, from this story came the saying '*Hinnaa fi Finna*', which literally means *Hinnaa* and kinship and '*Hinnaa fi Finiina*', meaning *Hinnaa* and the radiance of fire. The former attributes to the blood ties between the people with the name and the rest of the society. The latter is clearer in the sense that it connects the group with the *Hinna* prefix to fire. As a result, further sayings became popular among the society such as *tumtuu Hinna hinqabnee fi dubbii lafee hinqabnee tokko*, which means that any *tumtuu* group without *Hinna* is equal to discourse without substance.

67 See Knutson, K.E. 1967. *Authority and Change: A study of the Kallu institution among the Macha Galla of Ethiopia*. Gotenborg, Ethografiska Museot. Q(K)alluu as an institution and its attributes is in detail addressed by Knutson. See also Burka 2006.

6.2.2. *KUN ITTIIN BULTIIKEE HAATA'UU*: 'LET THIS BE ON WHICH YOU LIVE'

The second tradition on the emergence of the *tumtuu* group traces it to an incident that happened during an age group rite of passage. It holds that during the customary group circumcision of equal ages (peers), the 'father', who could be a godfather should provide gifts to each of the initiated and blesses them with the gifts upon which their future will depend. He is known as *mata Qabduu*, literally means 'head holder'. He was the one who sits behind the boy who is being circumcised and the one who covers the boy's face with a cloth. After the circumcision, the '*mata qabduu*' shall offer gifts to the initiated. According to the story, the 'father' distributed gifts to each of the initiated when finally he realized that he did not have any gift for the last person being initiated. Therefore, he took up the knife (*Qaraba*) with which the boy was circumcised and presented it to him with the blessing that '*kun ittiin bultiikee haata'uu*' meaning 'let this be upon which you live'.

Accordingly, the person who became the *tumtuu* was '*kan haaduun irratti hafe*' literally 'on whom the knife was left'. In other words, the *tumtuu* was the last person in the group to be circumcised and who took the circumcision tool to home. The question remains why the smith is attributed to the last to be circumcised. What is the metaphorical significance of the story? Does it hint that the *tumtuu* were those entitled with the 'left over' or indicating their lower status or those adopted by the Oromo? What is common among the Oromo social engagement is the fact that seniority upholds the right to access to something. It is possible that despite equal ages those from senior lineages were those who had right to be circumcised first.

It is common to hear in the form of a rebuttal from the craft group against members of the mainstream society who ridicule them by saying if *it were not for us you could remain hanging onto your mother*. This implies more than just that they were the ones who produce circumcision tools. It rather refers also to the fact that it is members of the *tumtuu* who provide the role of midwifery, circumcision etc. The mainstream Oromo regard the role as polluting job. The ethnographic observation from Wollega has provided an understanding of the crucial role of the *tumtuu* (smiths) in various facets of the society. In other words, the *tumtuu* are those who not only provide tools for utilitarian purposes but also various ritual objects including amulets and slag for medicinal values

6.2.3. *KAN AMMA QABATTANIIN BULAA*: LET YOU LIVE ON WHAT YOU HAVE IN YOUR HANDS

The mainstream society (including former smelters) in southwestern Wollega, have this story, which is also popular among the artisans. The fact that former smelters, as well as the mainstream, claim that their ancestors cheated on the ancestors of the current *tumtuu* shows the attempt to distinguish between smelting and smtihing. The attempt to align the smelters with mainstream society implies the rejection of the group to be regarded as artisans (*cf. chapter II*).This tradition is particularly concerned with the justification of how the present *tumtuu* became what they are or how a new law changed their former status. In this case, therefore, the story does not deal with the origin of the first *tumtuu*.

On the contrary, the story points to the presence of various craft specialists comprising of the society at a given level or territory in Wollega. This story was about the transfer of responsibility from one group of the society to another section of it. The following is how both groups present the story.

According to the tradition, in the remote past, the Oromo ancestors had the habit of moving from one region to another. This spatial movement was in accordance with the rule that governs the settlement pattern of *Hangafa-Quxusu* relationship (for this relationship see Kassam 1999; Kassam and Megerssa 1989; Megerssa 1994 also Burka 2006 and *cf. chapter II*). This journey was said to have been manned by the Bull (*Korma*) that leads the livestock. Where the bull decided to stay, the Oromo settled either temporarily or permanently.

The story continues to hold that the Oromo of Wollega continued their journey in search of a suitable region to occupy. On their way to the present southwest Wollega, each of them was obliged to help one another carrying various objects in their possession. It acknowledges the presence of various craft specialists with their working tools in the company of the groups. Then, they came to a big river. The *tumtuu* asked the non-*tumtuu* man to carry his working tools across the river. Since it was part of the obligations, he received from him. A new law was 'beaten' after the people crossed the river. The new law after the river has changed the status of each side.

Abbaa seeraa, father of the law, declared that *'kan amma qabattaniin bulaa'* meaning 'make your livelihood with objects in your hands.' Everyone should respect the law. In this way, the occupational identity of the present *tumtuu* was forged. In other words, the present *tumtuu* members previously belonged to the mainstream society that includes smelters. Actually, this story is in contrast to the saying *tumtuu* without *Hinna* is the same as discourse without substance. Many including members of the *tumtuu* group claim that the main source of *tumtuu* identity is from lineages with prefix (H) inna. Consequently, the other saying *Hinnaa fi finiina*, which means *Hinna* and radiance, sparkle of fire underlies that it is from these groups that artisans who could work on metal forging originated.

In sum, the three stories I presented above convey message on the nature of stratification among the Oromo of Wollega. However, they are not enough to explain why the mainstream regards the *tumtuu* as low among the Oromo ranks. To redress this gap, I present other stories common in different parts of Wollega among both the mainstream society and the *tumtuu*. The next stories are not how the various occupational identities originated among the Oromo or how the Oromo society was stratified into different occupational and social identities. In contrast, the stories attribute to various incidents that caused the emergence of the marginalized position of the *tumtuu vis-à-vis* the rest of the dominant groups.

6.3. TRADITIONS ON THE ORIGIN OF MARGINALIZATION OF THE *TUMTUU*

Before I embark on the discussion of the marginalization of the craft people particularly the smiths (the *tumtuu*) in the Oromo society, a number of questions require explanation. The first I would like to pose is whether the *tumtuu* really were marginalized among the Oromo. This question requires further explanation as to why the *tumtuu* were marginalized and who were the marginalizing groups. Were the *tumtuu* or the non-*tumtuu* who marginalized the other? What were the attributes of marginalization? Was equal commensality permitted between the smiths and members of the mainstream? Was exogamy, for example, possible crossing social boundaries? If there was marginalization, was it strong or lenient? Was the mainstream equally marginalized all craft specialists (smiths, smelters, potters, tanners) or there is marginalization among the marginalized? To put it differently, was there any leniency on one and strictness on others? Was marginalization explainable from space utilization in the environment? Both the artisans and the rest of the population tell the following stories. We should note, however, that in the 'rest of population' are there smelters who reject being categorized into the group of artisans.

6.3.1. CONFRONTATION AT A FEAST: DISAVOWING THE KNIFE

After the ancestors of the mainstream society cheated on the ancestors of the *tumtuu* while crossing the river, the story holds that, both lived together without any confrontation. According to the mainstream's storytellers, the *tumtuu* participated equally in the community life of the society. They related that both enjoyed equal commensality in different aspects. They invited one another for respective feasts. The number of the craft groups in one place was also large. They thrived and flourished with the rest of the society.

Once upon a time, a feast was organized by the *tumtuu* for which the non-*tumtuu* members were also invited. It was a big feast. For the respect of each group (which can be according to lineages), fattened calves were presented for slaughter and feasting. Participants from the mainstream were also given one. When the non-*tumtuu* people asked for knives, the *tumtuu* responded indifferently. They replied 'go and make it for yourselves'. The non-*tumtuu* told them that they did not know how to make it. The *tumtuu* said they could learn to forge one. The non-*tumtuu*[68] (which they phrased it as 'we' including the smelters) were confused, ashamed and annoyed by such deeds of the *tumtuu*.

Thus, they left the feast area of the *tumtuu* and convened to discuss the matter. Particularly two issues worried members of the mainstream. These were the huge number of the *tumtuu* in single locality and the possibility that they would refuse to make weapons for them. Even worse was that they would revert to making weapons for their enemies. They seriously addressed these issues during the meeting of the non-*tumtuu* and a decision was made. Then, they decided to execute a concerted attack on the *tumtuu* groups elsewhere in

68 The story tellers (e.g. *Obbo* Oliiqaa Lamuu, *Obbo* Ejjetaa Jobir) present it as if it were very recent past and seem to still holding grudges towards the *tumtuu*. Part of the interesting issue in this story is that it was told by the mainstream including the smelters. For instance, *Obbo* Oliqaa Lamu belongs to the family of the renowned iron smelters. He puts himself in the category of the mainstream. *Obbo* Ejjeta Joobir is not smelter but shares the same 'we' with *Obbo* Oliiqaa. When I asked *Obbo* Ejjetaa, were you a smelter, he replied 'no' and added that smelters were 'people like us' not like 'them'- the *tumtuu*.

the region. In addition, members agreed that the number of the *tumtuu* should be kept very small and that they should not be concentrated in one area. As a result, the *tumtuu* were disbanded from their core areas.

In addition, various clans all over the Wollega region confiscated the members of the *tumtuu* among themselves. According to this story by members from the mainstream society including the smelters, it was since then that the *tumtuu* were kept low in status and denied equal commensality with members of the mainstream. Nevertheless, the *tumtuu* do not recognize the reason for their separation and 'marginalization' by the rest of the population. The following is an alternative story told by the *tumtuu*.

6.3.2. THE CONSPIRACY TO KILL A 'NOTORIOUS' CHIEF

This story is the one offered by the *tumtuu* as a reason for their 'marginalization'. On the extreme side, some members of the *tumtuu* claim that they were the ones who refused to mix with the rest of the population. Despite this, the story of the chief is very strong in the Yamalagii lineage of the *tumtuu* groups. According to them in the remote past, a chief ruled over nine lineages. He belonged to one of the nine lineages. The lineage to which the chief belonged was known for its warring and hunting skills. This skill of the chief's lineage has intimidated and kept them under the chief. The chief was notorious and cruel for which the rest of the subjects disliked him. The eight lineages began to conspire on how to eliminate the chief.

The suitable time when they were waiting for arrived during a seasonal hunting of big game. This game involved able-bodied men who could campaign far away from the residence. The chief remained at home with the females, old men, and children. The representatives of the eight lineages decided to assassinate the chief during a traditional checker game by sand gravel, called *sadeeqa*. The procedure in the game involved covering the face of one of the parties when the opponent was reshuffling *saddeeqaa*. The arrangement was to stab the chief during his turn when he was covered with clothes. One of the eight representatives of the lineages was a *tumtuu* man from Yamalagii lineage.

Accordingly, they ordered themselves into eight. In this arrangement, if the first person failed during the first chance, the second was to execute...etc. The arrangement put, the *tumtuu* representative as the last option. According to informants, the decision to put the *tumtuu* at the bottom of the order was due to the rest of the lineages' low regard for the courage of the group. The story concludes, however, that the lineage representatives from the first to the seventh were afraid to execute the assassination. Then, the turn of the *tumtuu* representative arrived. The *tumtuu* person who was regarded a coward could stab the chief to death.

However, the death of the chief particularly in the hands of the *tumtuu* had grave consequences for the Yamalagii lineage and their *tumtuu* members. Following the news of the death of the chief, members of his lineage who were on campaign were summoned back to the village. They found that the *tumtuu* of the Yamalagii lineage had killed the chief. The chief's lineage waged war on the rest of the lineages who were already on the run. They in particular dealt with the members of the Yamalagii lineage to which the *tumtuu* killer belonged. Since then, the *tumtuu* remained under control and barred from commensality with the rest of the population. Their number also remained very small. In addition, they were no more cattle herders unlike their predecessors.

6.3.3. REMARKS ON THE MARGINALIZATION OF THE *TUMTUU*

The type of social organization in which it operates can characterize the nature of distancing from commensality with the mainstream. It is apparently understood that in highly centralized traditional societies as opposed to the less structured ones, the position of the artisans proportionally vary. Accordingly, in the former exist separate occupational identities and in the latter many groups merge into single caste (Haaland *et al* 2004a). In both Abee-Dongoro and Aira-Gulliso sites, iron producers remained within the buffer zones of the various competing clans. This case particularly differs from areas under hierarchically structured such as the Omotic-speaking societies (Haaland *et al* 2004a, 2004b; Haberland 1959; Todd 1977, 1978).

Among the Oromo, 'members of the mainstream society' have conducted iron smelting. It is possible to argue along the lines of Haaland *et al* (2004a) that the level of structure of the society might have determined the categorization of various occupational identities in Wollega. The fact that one occupational identity called *tumtuu* constituted about three occupations shows that there was indulgence on the nature of marginalization of craft people among the Oromo society of Wollega. This is, however, despite the presence of ranking and separate identities.

Furthermore, its smelters dictated the main source of iron for regional and international markets. Thus, since the smelters had backing from the body of their respective lineages, the issue of marginalization is concerning only the *tumtuu* occupational groups who depended on the smelters for the source of the material. The *tumtuu* (smiths, weavers, and potters) is an endogamous group, forming a separate occupational identity immediately above the tanners in the social ranking (cf. *chapter II*).

Although there is no distinct way of life (in house structure, clothing and decoration or physical difference (e.g. see Shack 1964 for the Fuga among the Gurage) to differentiate them from the mainstream Oromo, they mainly inhabit in separate hamlets in the study areas. Separation of habitation sites is one major visible distinction from the rest of the population. Nevertheless, there is no clear geographic boundary set between the *tumtuu* and the non-*tumtuu* communities. The presence of habitation sites mixing smelters and smiths further complicates the attempt to distinguish between the marginalized and the non-marginalized groups.

Among the mainstream Oromo society of Wollega, it is common to observe the distinction between the smiths (*tumtuu*) and tanners (*duugduu*) through some metaphoric terms. The *tumtuu* are associated with dryness (*gogeettii*) while the *dugduu* with wetness (*jiituu*). I would like to note here that the 'dry' metaphor is addressed to the smiths than the weavers or potters. Unfortunately, however, since it is difficult to distinguish the weavers from the smiths as both occupations can be performed by individuals from the same family members and since the potters are sometimes[69] the wives of smiths, the attempt to attribute 'dryness' only to the smiths is very fragile.

In addition, it poses another question: who are the marginalized? Were they the smiths? Were they weavers or potters? At a concrete level, it is possible to assume that the ascription of 'dryness and wetness' originated from the materials both groups dealt with-metals and hides. Metals are regarded as cold, lifeless, and dry. Hides are still wet although the Oromo term for hides connotes equal meaning of 'dry' namely *gogaa*.

However, rather than associating the ascription with their materials, the mainstream society went to the extent of group's natural behavior. It relates, for instance to bad or good omen while encountering with members of these groups. The manifestations of such behaviors were believed to occur during face-to-face encounters with them on streets. If a person destined to attend some appointment early in the morning, met a *tumtuu* man or woman, he 'knew' that it was a bad luck. As a result, he should cancel the appointment and return home. This is defined as 'bad omen', *milkii yaraa*.

On the contrary, when one meets a *duugduu* early in the morning (before meeting somebody), it is regarded as 'good omen', *milkii gaarii*. He knew that he could proceed to attend the affair. It is similar to what scholars distinguish between two types of polluters: 'sterile' and 'fertile' polluters (Freeman 2001; Pankhurst 2001). Despite attributing the tanners to 'good fortune', they are far below on the margin of the society than the *tumtuu*. They are regarded as people possessed by evil spirits and with evil eyes. The extreme saying *'gabaan duugduu fi gabaan saree harbuu jala'*, which means 'the market of tanners and dogs is under an oak tree' explains their categorization with dogs who live on scraps from the skins and hides and the physical distance from other marketers.

There are a number of explanations forwarded for regarding the forgers close to the chiefly class, the Boorana. In addition, there are different ways of showing that the forgers are regarded as despised, which was in the past prevalent among the Oromo of Wollega. One of this is with regard to sexual relationship between the smiths and their counter parts. The main question is not the common one i.e. whether exogamy was allowed or not. It is rather, whether sexual partnership (even at a temporary level) permitted. The question goes on to clarify whether there is a 'healing' medicine for crossing the socially and supernaturally closed borders due to sexual intercourse.

According to informants, in the past, members of the mainstream considered it dangerous to take forgers as sexual partners. It is also rare for forgers to aspire for sexual or marriage relationship with the mainstream including the smelters. It is because the *tumtuu* 'feel' that the mainstream does not take them with equal sexual commensality and as a result regarded crossing the border as costly adventure. The mainstream Oromo forward a number of reasons for regarding the forgers as lower, impure and, therefore, should keep

---

69 Unlike in some regions, in northeast Wollega of Abee-Dongoroo district the wives of the smiths are not really working on pottery. They are house-wives. Nevertheless, in southwest Wollega of Aira-Gulliso, some of the wives of the smiths still maintain pottery crafting.

the distance between these different identities. Their impurity is probably associated with their connection with deadly/dangerous materials, iron, and fire. Due to their association with fire and their 'sexual dissoluteness', the forgers are believed to 'burn' their partner, *tumtuun nama gubdi*, and cause impotence or even worse might cause death.

Therefore, disguised in such ascriptions, the *tumtuu* and the rest of the society were made to maintain their separate identities. For the question I posed to a *tumtuu* in Wollega to explain what mainly distinguish them from the core of the Oromo, he replied that it was circumcision. Since, today, all the Oromo male and female are circumcised, the *tumtuu's* reply seems baseless. Nevertheless, the *tumtuu* insisted on the story passed to them from their parents that holds that only the *tumtuu* members of the Oromo were circumcised in the remote past. The informant was arguing that the *tumtuu* were who refused to marry outside their group, not the other way round.

This is probably a rebuttal on the part of the marginalized artisans. However, it is true, informants from the mainstream agreed that the Oromo chiefly class of the *boorana* were circumcised late during adulthood. Information on precise age is lacking but some attribute the period to the liminal time to leadership in the Gada Assembly (40-48 years old). According to this, the Oromo were circumcised around age 39 years before entering into the fifth cycle at the age of 40 (Legesse 1972).The significance of the age is also that it was immediately afterwards that marriage took place. Since (for example in Wollega) the *tumtuu* did not participate in these age grade initiation ceremonies (e.g. Hultin 1984; Knutsen 1967), except probably performing some ritual roles, it could not have been obligatory on the part of the *tumtuu* members to wait until 39 years old to be circumcised.

This is similar to observation by Haaland *et al* (2004b:158) that when restrictions on the caste groups are by passed, some rituals were needed to avoid supernatural inflictions and social sanctions that would follow. From this perspective, I shall discuss how some rituals were needed in order to avoid supernatural sanctions brought because of pollution due to interactions. For instance, if a non-*tumtuu* person passes a night with a *tumtuu* woman in informal sexual relationship, he should perform some interdictions to protect himself from damage on his fertility and from bad luck. This ritual practice involves anointing ones seat with cow dung. The significance is said to 'cool down' the 'fire' and this is put in Oromo language as *qabbaneeffachiisa*. Cow dung is a symbol of fertility and used during marriage ceremony. Accordingly, it is put in front of the doorstep on which the bride and groom step before entering the house. Change of the number of stone supports in the fireplace is also another means of appeasing sanctions by supernatural power.

In Wollega, when a non-*tumtuu* marries a *tumtuu* woman, since it is believed that death might hit those who breached the social restrictions or at least fertility would be denied, some ritual interventions are performed to alleviate the problem. In a normal marriage, the wife builds a fire place with three stone supporters, *sunsummaa*. However, when a woman from the *tumtuu* is married to a man from the mainstream, she has to increase the number of stones in the fireplace.

Indirect information from such a story is about who was able to breach the sanctions laid down to maintain separate identities. It is clear from the story that the *tumtuu* were regarded less marginal to the mainstream society. In other words, the question for the sexual partnership seems to come only from the side of mainstream society. There is no any reference to such prohibitions and restriction in case a *tumtuu* man marries a non-*tumtuu* woman from the mainstream. The question is whether this has originated from the belief that the *tumtuu* burns her sexual partner or not?

The fact that the forgers are said to burn their partners because of their connection with fire can be challenged from two perspectives. The first is that only the males are associated with fire though female *tumtuu* who work as potters also burn their pottery. Thus, there is no reason to treat female *tumtuu* from this perspective as not all *tumtuu* female work on pottery. The second problem is that the term is an all-encompassing to include not only the forgers but also the weavers and the potters. The work of the weavers in particular does not have any association with fire.

However, it is possible that the dangerous position given to the forgers originated from their association with fire. Fire like iron has dual meanings in Oromo. Fire is a source of life but is also considered as destructive and polluting. Its symbol of fertility is reflected in that when an Oromo elder blesses married couple, he says *ibiddikeessan haa gabbatu,* which literally means 'let your fire prosper'. In this case, the man is referring to body thermodynamics created due to sexual intercourse. He is therefore wishing them to produce children. When somebody's wife dies, an Oromo has the saying '*ibidasaatu dhaame*', meaning 'his fire has gone out'.

According to this, the man's fire is his wife and probably vice versa. The wife therefore symbolizes life or source of life, which in turn means that fire symbolizes life. Fire is a medium of transformation of things from one state to another state-for instance from natural to cultural, from uncooked to cooked-a transformation, which is not reversible. The fire between man and woman as well creates a transformation to a fetus in the womb of the woman- the transformation to life, which is not reversible either.

On the contrary, fire is considered dangerous because of its powers of transformation and transmission of things. Fireplace, which serves as a medium of interaction can become medium of pollution. For instance, it is not advisable to share fireplace with such persons as infected with sexually transmitted diseases as well as leprosy. The belief is that fire can transmit the disease from the victim to the rest of the attendees. Fire is also feared as destructive element.

There is a saying among the Oromo that 'fire should be respected inside the fireplace'. It implies that if fire escapes from its controlled place, the damage it can bring could be huge. Such a belief might have been a basis for the categorization of the *tumtuu* as burning their sexual partners. The manipulation of fire by the *tumtuu* to transform one object from one state to another state should lend a ground for such a metaphor employed to describe them. The *tumtuu* also occupy ambiguous positions as fire does. They produce objects of production as well as objects of destruction.

However, according to informants, the wider distance, which was created between the mainstream Oromo and the *tumtuu,* has started to change since the socialist period in

Ethiopia in 1974 when members of the group began to obtain some rights such as owning plots of land as well as assigned in local administration. In addition, the expansion of Protestant Churches in different parts of Wollega (particularly in the northeast) in the last 20 years had important consequences on changes of perception particularly among the younger generation. Ethnographic observation in northeast Wollega where Protestant Christianity began to penetrate since the downfall of the Derg in 1991 revealed that in social and religious gatherings, the *tumtuu* have been largely integrated.

One serious impact that took place in the early 1980s was 'villagization' program of the Derg regime that has dissolved separate hamlets and mixed them together. However, since 1991 the project has dissolved with the new regime that led many to return to their former villages. As a result, today, the habitation of the *tumtuu* communities is markedly distinct from the rest of the population.

6.3.4. WHO OWNS THE KNOWLEDGE, THE GRANARY MAKER, OR THE POTTER? CONTRADICTING STORIES ON THE ORIGIN OF IRON SMELTING IN WOLLEGA:

One of the challenges regarding the study of iron smelting in Wollega is the contest between smelters and smiths on the ownership of the skill and its legitimacy. Such a challenge is evident in stories related to the origin of iron smelting particularly in northeast Wollega. One of the stories predominantly told by the group of smelters associates the discovery of the ore with a granary maker. The other story prevalent among the smiths attributes the first smelting to the potter's kiln. According to the smelters, once upon a time in the remote past, there was a woman named Balfee Saqoo, who has married to a man from the Malkee lineage. She was a granary maker and had a pit, *boolla gumbii.*[70]

A. 'The granary maker discovered it'

Balfee Saqoo was a poor woman who lived by making granary, *gumbii*. However, she was said to have been married to a man from the senior class of Malkee lineage. From the pit, she extracts the clay to mix with chaff and makes the container. The pit was a permanent place for granary making for the woman. Year by year the depth of the pit increased. One

70 ***Boolla gumbii***, a pit for granary making is one of the activity areas in the compounds of traditional houses in Wollega countryside. It is one of the activity domains of the women. It is so cultural that each of the house-wives should have one from which she prepares various sizes of sectioned granaries from the clay soil. Unless the soil becomes coarse and unworkable, the pit is always maintained. During the rainy time, it is filled with flood that she now and then reopened before the winter harvest time ensues. As the women are also participating in the harvest of crops in the field, they have to prepare the granary long before the crops were ripe immediately after the rainy season was over. Whereas the number of granaries in the living room symbolizes the wealth of the farmer, the quality and its finishing decorations are signs of the level of the skill of the house-wife. There are two types of the granary: the ones with a number of sections (separate one fitting one on top of the other) are called *gumbii* but there are other small size granaries without sections are called *gushgushaa*.

day, she realized that the aggregate of clay soil she was extracting has changed. It became very coarse and heavy. She sought to find a person who knew about this. One day an event took place that changed a history of the region.

There was foot track passing close to where she was working. When she saw a man pass by loading a thing (it was also an ore from far distance areas somewhere in the lowlands) on donkey's back, she asked him to examine the clay soil. The man was from the highland of Horro. He checked and asked her to take it with him and return with the results. Weeks later the man returned with a broad smile and told her, she has discovered a fortune. He has also brought her a sack of grain as a prize for her discovery. They call the mining site in northeast Wollega, Malkee's 'gold' site. It seems they have given high value to the ore. They do not call Balfee's 'gold' site, however.

Balfee was a woman married to a Malkee man. They explain in light of the Oromo local tradition of land tenure system. However, the mining site is now revered as ancestral sacred site as Balfee's sprit place. The ancestors of the Malkee lineage members inhabit surrounding the mining site at the locality of Faallee and used to make annual sacrifices to the site until in the 1970s. Therefore, they justify the reason for lineage members' sole responsibility and skill for smelting iron. Accordingly, members of the lineage through the Balfee's husband appropriated the right to the ore site and the skill to smelt it.

The story, however, lacks detail on how members of the Malkee lineage (cf. *chapter II*) learned the smelting following the discovery of the ore and the happy return of the obscure highlander.[71] It is not clear from the story whether the man was already a smelter or a merchant in ores. Furthermore, it is not clear whether some skill holders accompanied the man to begin smelting at Malkee.

B. 'A potter discovered it'

The second story is about how a potter discovered a smelted material in her pottery kiln, *boolla suphee*. This story is common among the smiths from the Yamalagii clan, who inhabit a separate hamlet from the Malkee lineages. According to this story, once upon a time in the remote past, there was a woman potter in the village, who was an ancestral mother of the present *tumtuu*. She belonged to a *Yamalagii* lineage. The potter had a pit (kiln) where she burnt her pottery. She always lined her pottery in the pit and covered it with acacia bark, *shollooqee laaftool*, and set fire onto it.

71 In May 2009, I have surveyed part of the highland Horro around an historical politico-ritual site of Odaa Bulluq believed to have been established around 15th century (according to the genealogy count of the descendants of the founder). One purpose of the fieldwork was to investigate one of the destinations of metal tools after they left the smithing site. During this, traditions collected about the source of iron in the highlands of Horro exclusively refer to the lowlands of Abee-Dongoroo. There is not any reference to the existence of smelting activity around the ritual sites where still survive descendants of smiths. Smiths recall that their parents used to obtain blooms and bars from Abee-Dongoroo to work onto various tools-both secular and ritual one.

One day, before she conducted another pottery burning, she was cleaning the bottom of the pit from previously accumulated ashes when she found a coagulated substance at the bottom of the pit. She took it out and showed it to her husband. The husband was a skilful person who knew how to make things. He tried the new substance and could make a thing out of it. Since then, according to the story, the person started to smelt iron. This story does not justify how they started smithing. Rather, it was about the source of the first smelted iron.

Apparently, therefore the smiths underline their entitlement to the knowledge of iron smelting to members of the clan. The intriguing side of this story is to know whether it implies that the knowledge was usurped from the present *tumtuu* by the Malkee lineage or whether in the beginning smelting and smithing were conducted by the same group. It is more intriguing whether the story also aims to articulate that the present smiths were previously among the respected members of the community as the smelters.

6.3.5. IRON AND ITS SYMBOLISM IN OROMO: PRODUCTIVE VS. DESTRUCTIVE POWERS

The Oromo belief that associates the origin of iron with their spiritual father from sky (Burka 2006, 2008, 2009; Kassam and Megersa 1989) is partly the background for the special ritual place the metal occupies in the society. Another myth of origin of the smiths in Wollega that traces to one of the first Oromo ancestors has also created a web of meaning around metals in general and in particular iron.

According to one myth of origin, the *Abbaa* Muudaa (Qaalluu), the Waata hunters (see Kassam and Megerssa 1989) with various body adornments discovered him in a jungle. The Waata are the 'blessed' hunters widespread in southern Ethiopia and northern Kenya among the Oromo (Kassam and Bashuna 2004). Among these adornments were iron bracelets on his arm and cowry shells. He was adorned in 'male and female' objects. In addition, he was wrapped into the *qomee*[72] cloth made in strips of black, red, and white.

It can be assumed that since black is related to Oromo supreme God, Waaqaa, it is a color of originality, purity, cleanness, and source of life (Megerssa 1994). This is evident in that the Oromo recognize the existence of two Waaqaa: Black and Red Waaqaa. According to them, both live in struggle of opposites. The former is the creator of all and likes to maintain everything in peaceful order. The latter is said to stand in complete negation as a force of destruction. As a result, the Oromo symbolize their creator, the black Waaqaa as *Waaqa-gurraacha garaa garbaa*, meaning Waaqa- who is as black as the belly of water.

The Oromo do not seem to have a primary term for the blue color. Consequently, water and sky are both black in color. In this case, one can posit that the reason behind associating iron with sky (black color in Oromo language) might have emanated from the high value

72 Qomee is a ritual cotton cloth worn during ritual ceremonies. It is weaved in tricolor: white background, red and black. The Oromo in general regard the tricolor as their national symbol-black is the color of Waaqaa, red represents the sacrifice people make in response of the promises of Waaqaa and white symbolizes cloud through which Waaqaa brings and performs his promises

they give to the material in comparison with other metals. In addition, this might have originated from their first experience with meteoric iron, which they understand it as metal created because of thunderbolt, a black one due to too much rusting.

However, the story that relates to the origin of the first *Kallacha* to have originated from sky or made by the first *tumtuu*, Hinno from the iron of Waaqaa gives the impression that there is practical reason to point to a celestial origin of iron. Ultimately, Waaqaa created both Hinno and iron. Hinno himself is regarded as the sparkle of Waaqaa, which is the same as the thunderbolt. Hinno forged *Kallacha,* an object of fertility from the iron of Waaqaa. It follows that thunderbolt is regarded as something fortuitous that brings fertility and prosperity for its owners. The distinction the Oromo make between iron and other metals is mainly one of color.

According to Wainwright (1942b), the Oromo vernacular term for metals (iron) is *siibiila* or *sibiila gurraacha*. He does not distinguish, however, whether the two Oromo terms were directed towards identifying iron from the rest of metals. From the Oromo vernacular term for iron *(sibiila gurraacha*), it might be possible to posit that the use of the term is to indicate its initial celestial source in the form of either meteorite or thunderbolt.

From this, an argument can be drawn that the term *sibiila gurraacha* (black metal) referred to as meteorite origin-and its ritual significance among the Oromo is because of its connection with what the Oromo consider high i.e. the sky and ultimately Waaqa, the supreme power. The high value accorded to this iron is the common belief that if one gets hold of it and places it in the granary or if one forges a plowshare from this iron and cultivates the land, it easily brings prosperity.

This connection between iron, the smith, granary, and agricultural produce symbolizes the fertility roles iron is attributed to play. I see this belief as related to the idea that iron will bring fertility to the people either in the form of agricultural fertility by tiling the land or human fertility through its symbolic intervention, as materialized in the phallic object, the *Kallacha*. As I have earlier pointed out, the origin of iron metallurgy in northeast Wollega and its association with the women working on the granary and pottery is closely connected to the idea that iron brings fecundity.

However, as well as the obscurity in its origin, metals in general and iron in particular occupies various positions in the Oromo society. Iron or metal is a source of life and a means to end it (Burka 2009). Products of iron change their meanings in accordance with the context (secular and spiritual) in which they are used. A hoe is generally a productive object in the context of farming. However, it becomes a symbolic weapon of death through 'tearing a body apart' in the context of Oromo traditional belief site, the *qaalluu*. Similarly, a harness is used to control and propel a horse or mule in the right direction. In this regard, it symbolizes good behavior. It is so powerful because it is put between the animal's jaws. However, its role of preventing the animal from chewing, for instance, becomes a symbolic weapon of curse to cast spell on somebody who did wrong. In this, thus, it becomes a tool for bringing dumbness, deafness, and ultimately death.

On the other hand, metal tools are used to protect persons from evil and acts of sorcery. One can illustrate this by looking at a time when a new house is built in the Wollega countryside.

Prior to the building of the wall, five species of wood, which are tied together with 'milky' climbers are erected where the 'backdoor' is supposed to be located. The procedure is known as *shanan dhaabuu*, erecting the five. It is the symbolic door. According to local tradition, a broken axe and slag pieces are buried under the symbolic door including observing other ritual sanctions (Burka 2009). The axe and slag are medicines for protecting the house from envious neighbors who could cast spell during the critical time of putting things for building. The story offers an important insight to archaeologists who work on habitation sites.

There are also other critical problems where metal (iron) tools are used to alleviate or prevent a problem from happening. Metal tools or pieces of them are carried when people travel during the night. Darkness is regarded as the time when evil spirits are unleashed. Metal objects are also kept close to the bed where a woman has just delivered. This could be an axe, a knife, or a spear point. The belief surrounds that the woman who has just given birth is a likely to face visitations by bad sprits. It is period of vulnerability that needs to be observed.

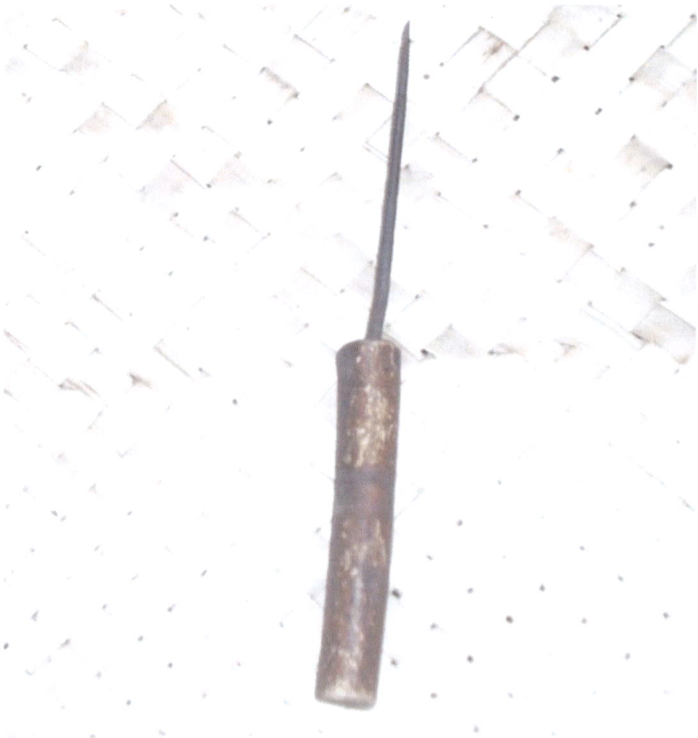

Figure 67 a wooden hafted needle, mutaa, amongst women's belongings in rural Wollega

Metal tools are sought to enable her to pass this critical stage. Similarly, new born infants, calves or donkeys are believed to likely face disturbances from evil forces. Metals are, therefore, used to protect them from unwanted interferences, in the form of cowbells. For instance, newly made girls' hair at puberty is decorated with a needle, called *mutaa* (Figure 67).

The changing contextual meaning and role of metal tools can very well be explained through a needle called *mutaa* (Figure 67). Under conventional circumstances, needle performs different functional roles such as sewing clothes or baskets and unmaking woman's hair. It is also used to take out thorns or other sharp points that pierced one's feet. In addition, it is also used as a marriage gift in a ceremony known as *suuqqata*, where some ritual evergreen grasses (*chooqorsa*) and other objects are together inserted into the hair of the would be wife. Therefore, its economic and social value is very significant. At other times, this object with other tools made of iron, is used to protect beasts from attacking domestic animals by inserting it in the fireplace and whispering prayers (Burka 2006, 2009).

Despite such practical and symbolically productive roles it plays, *mutaa* can also be used as instrument of destruction and death. For example, *mutaa* is used as an object of intercession between people who faced enmity. It becomes an object of curse and oath. Sometimes,

Figure 68 a sacred site, siidaa of the Odaa Bulluq in northeastern Wollega. Note that the evergreen podocarpus tree is encircled with stele in the compound of which and through of which various rituals including conflict resolving are conducted.

Figure 69 A metal harnesses, saamsaa, used in the traditional ritual court to resolve conflicts

compatriots might betray each other on their common business. There might also be an occasion when someone suspects another person of stealing his property for which he cannot produce tangible evidence. Even worse, there is time when the plaintiff could not bring the defendant to the spiritual site. Then the symbolical role of the *mutaa* would be interjected to resolve the complicated problem. The wronged would go to the spiritual site, discuss with the council of the spiritual head (see Burka 2009) and as a last resort inserts the needle into the revered site, the *Siidaa*[73] (Figure 68) or at the ritual fireplace.

However, when the defendant, who is worried by the spells, decides to appear at the ritual court, one of the metal tools exhibited by the councilors to judge the matter was the *mutaa*. In other words, the defendant would swear in the name of the symbolic powers of the needle. I will add some two more objects to elucidate how ordinary metal objects change

73 *Siidaa* is a revered compound, encirclement around a Podocarpus, an evergreen tree of Taxaceae family. It is a symbol of purity and beginning. It is called *booran* tree-meaning the first born, the eldest of all plants.

their meanings once they left the utilitarian purpose for which they were produced (see Figures 69 and 70).

One of the objects at the possession of the ritual sites in Wollega and employed to resolve conflicts through the rituality of the objects is called *saamsaa*, harness and bit (Figure 69). The pragmatic significance of this object is clear. It is used to control horses or mules. Since the harness is put between the jaws, the animal is prohibited from eating or chewing grass. It is also deterred from making undesired movements. It enables the man who mounts the horse/mule to easily propelling him to the desired direction. *Saamsaa* also prohibits the animal from making noises. *Saamsaa* as the Oromo meaning shows is an instrument of muffling. It is generally an instrument of suppression and control of the wild behavior of the horse.

This extraordinary power of the harness is metaphorically/symbolically transferred to the traditional ritual site of Wollega. At the ritual site context, *saamsaa* takes supernatural meaning. Symbolically, this object serves to harness wickedness and directs people's behavior to righteousness. In a similar procedure, the wrong doer begins to call curses on himself, other members of his family and his property. Some of the contents of the curse

Figure 70. the hoe, the metal harness, Qonyee, and the rest of other wooden components of the Ethiopian traditional plough

include the saying 'let *samsaa* of Waaqaa shuts my mouth, the mouth of my family, and my cattle'; 'let me or my family spell out no word again if I did this what I am accused of'.

The defendant would continue to say, 'Let the *saamsaa* of Waaqaa bring on me the dumbness that the object causes, let my mouth be shut'. Then the councilors of the ritual man would continue to bring more objects that the defendant would either jump over them or touch the sharp edges and call on himself all the possible damages the weapons bring.

The other subsets of objects of curse (Oath) include the two inseparable parts in plow farming-the hoe and its harness, *Qonyee* (Figure 70). Their primary use is very clear. Whereas the hoe is used for splitting the ground, *Qonyee* serves the role of harnessing and griping the rest of the implements into one set. It is because *Qonyee* fastens the farming components together that the hoe splits the soil apart. The hoe cultivates land. It penetrates the 'mother earth' and prepares the farm for cropping. It is an instrument of agricultural production and fertility. It is a symbol of manhood. To own it means to be able to produce and feed oneself or ones family.

Such primary pragmatic characteristics of the tool are used at the ritual place occupying the meaning as an object of punishment. To change the primary characteristics of the materials, all metals of curse or oath are dipped into the blood of an animal and are annually consecrated in blood. Thus, blood changes their meaning. Under the circumstance when the 'wronged and the wrong doer' present their cases to the ritual court, this set of materials are brought in front of the judges. The ritual father and his councilors (called *shanee*-the five and *salgee*-the nine) use this object to rule over the matter and as a truth finder. This becomes imperative under the situation when the defendant denies the alleged crime. Then the ritual heads ask if he would like to take an oath on the object. If he agrees, then he has to pass through all objects in the package and gets his freedom. According to members of the *Qaalluu* councilors, however, they have observed during many occasions when defendants start taking oath but quit trembling from fear of the spells that can be brought on them.

The accused would echo after the head of the *Qaalluu*'s *shanee.*[74] According to the spiritual father, (*Qaalluu*) *Obbo* Abdisa, the wrong doer would say if I did this or that let the sharp plough of Waaqaa penetrate my body, my kids, and my family. The accused would continue to say let the strong grips of the *Qonyee* of Waaqaa make me immobile, paralyze me. He also says let my farmstead remain uncultivated, let me not have any produce that *Qottuu* and *Qonyee* bring for life; let me, my kids and family miss the fruits of this object and be starved to death etc,. The oath does not stop here. It continues in the medium of other objects. However, it can stop in the middle of the prosecution if the accused decides to submit to the symbolic power of iron objects.

74 According to informants, the term *shanee* is a symbolical number to represent the physiology of a person (head and the four limbs). The same members of the council also become nine-called *salgee* representing the sacred, the nine Boorana. Nine members of the council are composed of five male representatives and four female representatives. Each of the group has their own heads. The head of the male group is the protector of the metal ritual objects particularly the Ark like object called *Kallacha* whereas the female head is in charge of the female objects as *chaachuu* (objects decorated with shells).

As I have pointed out in different parts of this work, the archaeology of iron production is not only concerned with the physical aspects of the objects. Rather, its focus should integrate the socio-cultural aspects of the trade. In this case, the context under which the object is found needs rigorous investigation. It is not enough that the archaeometallurgist finds and documents vestiges of ancient metallurgy. This ethnoarchaeometallurgical study lends a peculiar perspective as well as a caution for the archaeometallurgist on the changing meanings of an object with changing contexts. As we have seen under previous section, the same object that can be used for day-to-day utilitarian purpose can carry supernatural meaning when the context of its use is changed. It is particularly important to note the conflict resolution roles played by metal objects in the context of the sacred places of the Qaalluu (Burka 2009). Be it that as it may, in the following chapter, I will attempt to situate the Wollega material in its wider context.

CHAPTER SEVEN

7.0. THE ARCHAEOLOGY OF IRON PRODUCTION IN ETHIOPIA: A SYNTHESIS OF THE WOLLEGA MATERIAL INTO THE WIDER CONTEXT

...when the 'beaker folk' were conceived of as roaming about in Europe, it was within the continent of Europe itself.... [However], any innovations within the continent of Africa were only considered to be derived from people migrating to Africa from elsewhere. It was only later, in relation to... [for instance] the spread of the Bantu languages, that an intra-continental frame of reference was established...[but]still within a migrationist perspective. (Adapted from Sinclair, Shaw and Andah, 1993:2)

7.1. INTRODUCTION

The conjecture that has long attempted to frame the African prehistoric past as an impasse in human evolutionary developments (see Chirikure 2005; Holl 1997, 2008; Okafor 1993; Sheppard 2002; Trigger 1990) is well summarized by Sinclair *et al* (1993: 2) in their provocative observations. A number of intriguing perspectives can instantly be drawn from their assertion. On the one hand, it would include scholars who inclined to conclude that Africans have always been on the receiving side even back in the past. It gave them an impression that Africans always awaited any external impetus to awaken them from their deep-seated technological gradualism. This can be framed as for example the 'introduction of iron metallurgy on the head of stone using culture' (see Killick 1999, 2009; Miller and Van der Merwe 1994; Phillipson 1985, 2005; Pleiner 1980, 2000; Tylecote 1975; Van der Merwe 1980).

On the other hand, it puzzled and attracted critics who as a result carried the discussion in an antagonistic way asking why it is only particularly with the discourse on the African past that an outside origin of civilization was sought (see Alpern 2005; Chirikure 2005; Holl 1997, 2008; Sheppard 2002). The conundrum in the study of the African past particularly metallurgy becomes more intriguing when one finds that there is no such antagonistic debate on the source of the knowledge of metallurgy in Europe[75] (see Burka 2006; see also Pleiner 1980, 2000; Tylecote 1980; Van der Merwe 1980). Europeans seem to have acknowledged Near eastern source of the knowledge.

Thus, the question is why should there be such a debate in Africa. This chapter in particular is far from dipping into the hot debate and being a party to either of the

75 As far as I am aware of there is no division among scholars posing antagonistic standpoints on the direction of the spread of the knowledge. In other words, there is no such argument whether north Europeans were the originators or receivers as well as the south. A consensus seems to dwell on its source from the Near East.

antagonistic advocates of the origin of the knowledge of metallurgy (cf. *chapter I*). It is however inescapable in that the interpretation of African metallurgy (particularly in sub-Saharan Africa as well as in Ethiopia) is not immune from such a debate. In addition, it has constituted an example where research questions being posed by archaeologists and the interpretation of the outputs would affect the understanding of our enquiry into the matter (see Childs and Herbert 2005; also Sinclair *et al* 1993: 9-12). In other words, the question of origin, throughout the history of the archaeometallurgical research has been the major preoccupation of scholars in African. The debate on its origin and spread is still far from settled (see Holl 2008; Killick 2009).

This dissertation has sought to shed light on the archaeology of iron production in Ethiopia by 'reconstructing' past activities and reversing the chronology backwards. In other words, it is an ethnoarchaeometallurgical research meant to serve as interpretative model for similar activities in the past. The research is unique at least on three grounds. The first and foremost is its uses of a *Third Model* on the discourse regarding the culture-history of Africa's past, and is in this particular case, the archaeometallurgy of Ethiopia and the Horn (*cf. chapter I*). By implication, the discussion regarding aspects of iron metallurgy in Wollega is made possible by allowing the significance of both internal and external dynamics. The second distinguishing factor of this research is its attempt to single out and concentrate on a hitherto neglected aspect of archaeological enquiry i.e. the archaeology of craft specialization-in metallurgy in Ethiopia and the Horn. This is unique with the presence of ethnoarchaeometallurgical evidence to allow the observation of extant behavior integrated with production and use activities of metallurgy in Wollega. The third but still significant aspect is the regional focus and the approach of the research that attempts to address metallurgy as part of the culture whole. It is the study of the marginalized both spatially and topically.

Spatially, it focuses on Wollega, one of the regions left in the backwaters of archaeological study in Ethiopia and the Horn. As I have shown elsewhere in this dissertation, Wollega has proved an important arena to provide valuable source for regional and interregional comparison in iron metallurgy. It has offered its peculiarities. It is unfortunate that its potential was disregarded in Ethiopian archaeology. Disregarding its potential might have resulted from regional bias in Ethiopian archaeological research. Wollega has never been regarded 'archaeological' region during the century old 'modern' archaeological study in Ethiopia and the Horn.

Ethnoarchaeometallurgical approach is another thing, which is given here holistic approach to allow us reconstruct how people interacted among themselves and with their environment in Wollega in the medium of metallurgy in the past. Since 'the past is another country' (see Holl 2008) about which we know very little, ethnoarchaeometallurgy provides the experience of the present for interpretation of the past. This is where the *chaîne opératoire* framework has been unleashed for the archaeology of iron metallurgy to observe behavior in 'width and length'. The *chaîne opératoire* that enabled to integrate the human and the material world was instrumental to reveal the depth of the values given to metallurgy among the societies in Wollega (cf. chapters III &VI).

The emphasis on rituals for iron production represented by pronounced female anthropomorphic apparatuses, procreative terms, the connection between the myth of origin of the smith and the shaman as well as the stories of the technology's origin in connection with agricultural production and pottery tempts to conclude the antiquity of the knowledge in the society. In addition, it allows pondering on the reason why the smelters/ smiths innovated when they did. However, the question of why the smiths/smelters first started (if they did) remains to be a difficult question without further long term research in the future. Nevertheless, ample insight has already been obtained from this research.

The ethnoarchaeology of traditional iron metallurgy in Wollega is, therefore, one of the particularities to increase our understandings of the archaeometallurgy of Ethiopia. Furthermore, the research is timely in the sense that it has reconstructed and documented the one time dominant technological and socio-cultural aspects of the ways of life of the people of the region before the tradition has died out about half a century ago. This very well reflects Dave Todd's (1978) remarks that underlie the significance of the detailed inquiry of the parts in order to have a clear picture of the whole (cf. *chapter VI*). The cultural whole is the sum total of parts such as metallurgy. Metallurgy, here, is shown as mirroring the various ways of life in its medium including secular and spiritual aspects (cf. chapters III and VI). Wollega is part of the Ethiopian whole bridging different cultures and ethnoarchaeology is an approach(s) (see David and Krammer 2001) that can lead us towards gaining a picture of archaeological inquiry as a whole.

As we have already noted, a history of Ethiopian metallurgy is far from complete (Burka 2006, 2008, 2009; Haaland 2004; Phillipson 1993; Todd 1985). In other words, we are far from having the complete picture of the archaeology of metallurgy in Ethiopia. One can ask why. Was it because metallurgy was a new/recent appearance in the culture of the various parts of the people of the region? This is not the case since we know very well that metallurgy was known in the Ethiopian region and the Horn as far back as the middle of the first millennium BC (see Contenson 1981; Fattovich 1990; Phillipson 1977, 1993; Todd and Charles 1978b). Other intriguing questions may include whether metallurgy affected the life ways of the people of the region. Why are metallurgical studies in Ethiopia and the Horn different from the rest of sub-Saharan Africa, where the study dominates other aspects of the discipline or even 'more advanced than in many other parts of the world' (See Childs and Herbert 2005:276)? Was it because Ethiopian archaeology was the direct consequence of Egyptology /Oriental Studies that uphold the study of visible architectures and arts?

Before I embark on the synthesis of the data presented at various stages of the dissertation, I would like to forward arguments for the lack of focus on the archaeology of iron metallurgy in the Ethiopian region and the Horn (cf. chapter I). I begin with the premise that the problem is embedded in the way Ethiopian archaeological research has been structured and handled from the beginning. It is clear that modern academic inquiry into the archaeology of the Ethiopian region and the Horn started with the *Deutsche Axum Expedition* (DAE) as far back as 1906 (see Brandt 1997; Brandt and Fattovich 1990; Fattovich 1990; Phillipson 1990, 2005). This statement does not rule out previous separate sporadic attempts in different parts of the region (See Bent 1896; Brandt and Fattovich 1990) from as back as the 17th century (see Brandt and Fattovich 1990: 95). The problem I would like to attribute is the nature of paradigm (culture-historic) with which the origin of Ethiopian archaeology rooted (see Sinclair *et al* 1993 in the beginning of this chapter).

The question that hinges on the discussion of the beginning of academic inquiry into Ethiopian archaeology is how and by whom the research was handled in the course of time. For instance, what was the major focus of the DAE in 1906? Why was monumental architecture the 'epicenter' from which the interpretation of Ethiopian archaeological discourse springs? Why was the then Emperor of Ethiopia interested in the archaeological mission of what later became known as 'pre-Aksumite'[76] and Aksumite archaeology?

It is clear from the previous paragraph that the beginning of archaeological research in Ethiopia had a serious implication on the means and topical and regional focus of the research in the years to follow. It can be posited that archaeology has (directly or indirectly) been implicated in the political discourse of the Imperial period in Ethiopia. Did not Ethiopian royalty link their origin with the house of Solomon of Israel (see Tamrat 1972)? Were not Ethiopian Imperial regimes fascinated with the legend of the Queen of Sheba? Did archaeological enquiry at Aksum have any contribution to our knowledge of those skill-holders (masons, artisans, etc. and their tools) who made the civilization possible?

The preoccupation with elite architecture and monumental features (see Phillipson 1990, 1993) has to do with the interest that governed why the discipline was needed in the study of Ethiopian cultural past. Indeed, it was needed to reconstruct the 'root' of the Ethiopian rulers and legitimize their legendary origin, tracing it back to the Queen of Sheba (see Sellasie 1972; Tamrat 1972). Were not such the characteristics of African archaeological inquiry as far back as the *'savants'* who accompanied Napoleon Bonaparte's conquest of Egypt to the post-African independence period (see Shepherd 2002: 189; cf. *chapter I*)?

Thus, in the last over one hundred years, archaeological inquiry in Ethiopia strove to piece together bits and pieces of information on links between the country's elites and the outside world through reference to some superficial exotic finds (see Brandt and Fattovich 1990; Fattovich 1990, 2009; Phillipson 1990, 2009). As a result, after nearly about a century of research, renowned Ethiopicist archaeologists like David Phillipson (see 1990) have acknowledged that Ethiopian archaeology is still at its infancy, which is in particular true of the history of metallurgical studies (see Haaland 2004a; Phillipson 1993; Todd 1985). One can presume, from this, therefore that the archaeology of Ethiopia began in the service of the elites and failed to address the concerns of the ordinary members of the society (Phillipson 1993). Apparently, this explains why a history of Ethiopian metallurgy has not made much headway. In other words, a history of metallurgy would have a complete picture only when the archaeology of ordinary sections of society becomes central or at least part of the research endeavors.

In light of the above argument, I would like to emphasize that due to its control from the start by the royalty, archaeology failed to explore aspects of the lives and contributions of the 'common people'. Metallurgy and the people who own the knowledge occupied a marginal position among many different Ethiopian societies. Apparently, therefore, the *raison d'être*

76 See Fattovich 1990, 2009; Phillipson 1990, 2009 for the discussion on the beginning and problem of the use of the term 'Pre-Aksumite' culture for the period ranging throughout the first Millennium BC.

for leaving the study of craft archaeometallurgy in the backwaters of Ethiopian archaeology was embedded in the marginal position given to the skill holders of the trade. This coupled with archaeologists' preoccupation with tracing various attributes of ancient Ethiopia's civilization from outside (with some exceptions such as Francis Anfrey, Rodolfo Fattovich 1970s and 1990s etc.) has not allowed the archaeology of metallurgy to be studied on its own. This is well noted in Phillipson's (1990:55) statement that underlines the state of the art in Ethiopian archaeology (e.g. of Aksum) at the time, in the following clear manner:

> *Archaeology at Aksum has so far concentrated its attention on monumental architecture and on artifacts of foreign origin, which may be dated or otherwise understood by references to parallels from outside Ethiopia-often, [from] outside Africa... [Therefore] what we have ...is an archaeology of the Aksumite elite, represented by the residents and tombs of the rich and powerful who had access to luxury good.*

The bulk of the research works, which dealt with monumental architectural achievements of Ethiopia, did not seem to have noticed the significance of addressing the question of whether chisels made of metals or other materials were used to hammer them out of the gigantic rocks. Similarly, past research does not seem to have shown interest in the importance of understanding the identity of craft specialists and their socio-cultural basis. Since archaeology can help citizens build a better future by presenting well balanced research endeavors, the comments I underline here would be a reminder for the reassessment and reinterpretation as well as redirection of research focuses for the time to come. The challenge we observed above in the nature of archaeological research in Ethiopia is important in that it also affects the way we synthesize data from regions virgin to Ethiopian archaeology such as Wollega. First, however, I make a summary of the state of the art in the discourse regarding African metallurgy.

7.2. METALLURGY AND AFRICAN SOCIETIES: THE CASE OF NORTHEAST AFRICA ON ITS ORIGIN AND/ OR DIFFUSION

The advent of metallurgy into sub-Saharan Africa has long posed a serious challenge as opposed to the rest of the Old world (see De Barros 2000; Phillipson 2005; Vansina 2006). The major challenge lies in the direct transfer of population of the sub-continent from stone using cultures to 'metal ages' without pyrotechnological background (see Childs and Killick 1993; Killick 1999, 2004, 2009; Miller and Van der Merwe 1994; Pleiner 1980, 2000; Phillipson 1985, 2005; Tylecote 1976; Van der Merwe 1980). As I have highlighted in the introductory part of this chapter, one central theme that emerged in the course of the reconstruction of the African metallurgical past is the question of its origin (see Alpern 2005; Childs and Herbert 2005; Chirikure 2005; Herbert 2001; Holl 2008; Schmidt 2001; Vansina 2006). The question of origin has been the most hotly debated aspect of the study of African metallurgy (Childs and Herbert 2005; see also Killick 2009; Mcintosh 2005: 151-156; Mitchell 2005) and has negatively affected the nature of research.

The question of origin became an important theme in the study of African metallurgy because Africans were not considered to have been able to attain the sophisticated iron smelting technology without outside influence (see Chirikure 2005; Phillipson 1985, 2005; Vansina 2006). Thus, questions such as where did the technology come from? When did it arrive in sub-Saharan Africa regions? What were the possible routes? Who were the likely conduits or agents of the knowledge and many others (*cf. chapter I*) have become dominant and overshadowed other questions such as the implications (destructive or productive) of the knowledge on the African societies once it became integrated in the ways of life of the people.

Other researches concentrated on the 'ritual' and/or technological aspects of metallurgy in 'classical Iron Age' cultures (Childs and Herbert 2005; Killick 2004). The 'ritual' aspect of the study has been derived from ethnographic observations before the activity died out (see Cline 1937; Haberland 1959; Haaland *et al* 2004a, b; Todd 1985; Wainwright 1942a, b, c). The ritual category of iron smelting has been regarded as trivial and of no practical use (for details Barndon 2004; Childs and Killick 1993; Schmidt 1997; Van der Merwe and Avery 1987). This approach was, however, criticized for attempting to view the African setting by methods developed and extrapolated from a different cultural setting (Mapunda 2000; Schmidt 2001).

On the other hand, the focus on the technological aspect has particularly originated as part of the search for the conduits of its spread through the examination of physical attributes of metallurgical apparatus, for instance, the furnace (see Cleere 1972, 1997; Sutton 1985). Such choices of scholars to concentrate on the issue of the source of the knowledge (cf. *Chapter I*) have seriously obscured the contribution of the research for our scientific understanding (see Childs and Herbert 2005; Herbert 2001; Okafor 1993; Schmidt 2001). In other words, due to lack of focus on its contribution (productive or destructive) once the knowledge became integrated into the ways of the life of each society, the consequence of its introduction is left in the backwaters.

Archaeologists of African metallurgy continue to take different positions on the state of the art of the knowledge (Alpern 2005; Holl 2008; Killick 2009). Childs and Herbert (2005:277) have noted that the 'beginnings, development, and spread of metal using and metal producing cultures' are yet to be addressed. One can still add in parenthesis that the contribution of metals to the development of African society is yet to be resolved. On the other hand, some others favor the end of the hot debate on the origin of African metallurgy (see Alpern 2005). Stanley Alpern (2005:41) stated that the debate over the origin of African metallurgy was over. According to him, the debate ended in favor of those protagonists who supported the independent origin of the technology in sub-Saharan Africa. One wonders if this is truly the case (see Killick 2009). The fact that in recent conferences, sessions were organized to discuss the origin of metallurgy disproves Alpern's argument (for example see Holl 2008: 9). One can see Alpern's positions only from the vantage point of the fervent advocates of an independent origin of African metallurgy.

The following excerpt from Agustin Holl (2008: 10-11) highlights the state of the art of the discourse signaling that the dust of the struggle is far from fully settled (see also Childs and Herbert 2005; Chirikure 2005; Killick 2009; Mitchell 2005). 'Surprisingly' writes Holl '... no new development and/ or additional facts seem to alter the entrenched conviction that metallurgy was imported to and spread throughout Africa...by Carthaginians, Egyptians, or still unknown propagators in the middle of the first millennium BC....' This, according to Holl, is despite fresh data on sites associated with metallurgy in different parts of the continent substantiating earlier dates or at least associated with technologies unrelated to the so-called source areas.

I have previously discussed (cf. *chapter I*) that the radicalization of the scientific discourse denied African metallurgy a focus on the significance of the knowledge as an agent of change and development. It is well taken that each group is endowed with peculiar talents to contribute to the creation of a meaningfully functioning society. To do so, the society needs to organize its production activities. Those who are isolated are left to themselves, however. Change is the contradiction to the status quo. Change takes place through both internal and external elements in opposing struggles. The struggle and co-existence of opposites is a moving factor for change and development. This can happen through technological innovation (trial and error) that can possess either of three adoption strategies (Holl 2008:15). One of the options is to delay the adoption of innovations. The other is totally to reject the innovative ideas. The third option is to integrate or accept the innovation as quickly as possible after a certain period of trial and error. Innovation, as part of a response to internal challenges and as part of economic process, would possibly take place independently in different socio-cultural settings. Thi s would ultimately lead to specialization in various socio-economic and political activities (*cf. chapter VI*).

There is a view that craft specialization and organization of production provide the basis for changes in the society that ultimately result in the formation of ranks or chiefdoms or states (see Chirikure 2005; Childs and Killick 1993; Costin 2007). This might differ, however, according to controlling mechanisms and work schedules set in place for craft specialists. One could discern the difference between independent specialists and attached ones as well as full-time and seasonal craft workers (Chirikure 2007; Costin 2007). Metals and metalworkers occupy significant ideological, economic, cultural, and political places in various African societies. By implication, the holistic role of the technology could be seen as an impetus for specialization and control of the craft and its specialists (cf. chapter VI).

However, since no society is close to influences from its neighbors (negatively or positively) and is not virtually closed to interactions or contacts, the perspective to be adopted is more complicated than unitary explanation for changes to take place in the society (*cf. chapter V*). A glimpse at the regional geographical and socio-cultural setting of Ethiopia and the Horn, for instance, confirms the necessity of caution in addressing any kind of development in the sub-region (*cf. chapter I*). One major theme of this dissertation is the need to take the ***Middle*** position that benefits both protagonists of single/outside vs. multiple/inside origin assumptions of African metallurgy. The Ethiopian region and the Horn is central in this discourse in that it is regarded as one of the conduits for the transfer of the knowledge from southwest Asia (see Mapunda 1997; Miller and van der Merwe 1994; Phillipson 1993; Selasse 1972; Shinnie 1985; Todd and Charles 1978a, b; Trigger 1969; Tylecote 1975; Van der Merwe 1980).

Movements of people and materials have swept the region of Ethiopia and the Horn, due to its special geostrategic position connecting continents. The regional exchange network (*cf. chapter I*) had started with objects such as obsidian as far back as the seventh millennium BC. Then, it has culminated in the population movements, permanent settlements, and foundation of state-like structures in the middle of the first millennium BC in the northern highland regions of Ethiopia and the Horn. The development that took place in the northern Horn in the first millennium BC is similar to its counterparts on the other side of the Red Sea in southwest Asia (see Durrani 2005; Fattovich 1990, 1996, 1997a, 1997b, 2009; Mapunda 1997; Mitchell 2005; Munro-Hay 1993; Phillipson 2005, 2009; cf. *chapter I*). Thus, any endeavor to reconstruct the cultural past in the Ethiopian region and the Horn should take into consideration regional complexity implying that no single explanatory model is fully feasible.

A similar challenge is apparent in the discussion regarding the origin of the knowledge of metallurgy in Wollega. Wollega is a transitional zone through which various population movements, interactions and their ideas could have taken place in the prehistoric and historic past (*cf. chapter II*). The introduction and development of a cultural trait including metallurgy could be seen in light of the regional dynamics including population movements such as attributed to the Oromo back in the 16th century AD.

However, before alluding to the possible source of the knowledge of metallurgy in Wollega in particular and western Ethiopia in general, in the following section, I will piece together evidence from the two study areas of the region. Then, I go on to compare the Wollega material(s) with literary and ethnographic evidence with regions relatively close to Wollega (*cf. chapters II and I*). Finally, I will attempt to situate the ethnoarchaeometallurgical evidence from Wollega in light of our understanding of how the people in the region might have used the technology to deal with their environment since antiquity. In other words, the Wollega ethnoarchaeometallurgical material is sought to enable the formulation of an interpretative model in the study of archaeometallurgy in Ethiopia and the Horn.

7.3. EVIDENCE FROM WOLLEGA: TECHNOLOGICAL

Wollega, due to its geographical position in particular presents a number of opportunities and challenges for discussing its culture-history and that of the adjacent regions. Geographically, Wollega is situated on the crossroads between regions north of the Blue Nile where Semitic, Cushitic and Nilo-Saharan language speakers dominantly inhabit and areas westwards along the Sudanese border of the lowland regions where a Nilo-Saharan speaking population dominates. A number of rivers and streams originate from Wollega and traverse the region mainly as tributaries of the Blue Nile, which might have served as point of interaction and cultural development. Southwestwards across the Gibe River are the Oromo and their Omotic-speaking neighbors where various kingdoms have once thrived (see Abir 1966; Hassen 1990; Zewde 1991).

Thus, partly due to its geographical position across which population with their distinct cultures traversed, and partly due to its diverse climatic condition, which allows various ways of life, Wollega might have attracted population settlement in various periods since antiquity. Various cultural groups might have crossed and re-crossed the Blue Nile to roam throughout the region leaving their imprints in various forms. However, due to lack of archaeological research in the region, it is not yet possible to establish any spatial and temporal cultural sequences for Wollega and its adjacent regions.

The 'emergence' of the Oromo speakers in the 'sixteenth century AD' in Wollega must have resulted in much complex dynamics in terms of population and culture. The overwhelming presence of the Oromo in Wollega has today overshadowed what it would have been like before the '16th century'. Furthermore, the Oromo institutions of adoption, *Guddifachaa* and *Moggasaa* (cf. *chapter II*; Hassen 1990) have obscured the search for vestiges of 'remnant societies' in the region under discussion. Through these institutions, the Oromo have 'homogenized' the population of the region by integrating various identities beyond recognition. The most intriguing aspect of *Guddifachaa* and *Moggaasa* is the fact that the adopted individuals or groups would totally lose their primordial line and take the genealogy of the adopting groups or individuals. As a result, although some scholars argue that 'pre-Oromo remnants' (Gidada 2001; Hassen 1990; Waqweyya 1988) are discernible through the names of lineages and clans of the artisans (cf. *chapter II*), my probe into their genealogy and myth of origin (cf. *chapter VI*) points to their common origin with the Oromo (see Gidada 2001; Ta'a 1980; Triulzi 1996).

The ethnoarchaeometallurgical study of iron production in Wolllega has shown that the region provides unrivalled opportunity for filling the gap of knowledge on the regional culture-history in the past, particularly regarding metallurgy. The study served as an inspiring model for investigating whether technological variability in metallurgy points to the antiquity of the knowledge in the region (see Schmidt and Avery 1987; Schmidt and Childs 1985; Schmidt and Mapunda 1997; Schmidt 1997). It also showed whether it reflects the various ways of human adaptations to environmental challenges and opportunities (Alpern 2005; Chirikure 2005; Herbert 2001; Holl 2008; Mapunda 1997, 2000; Schmidt 1997, 2001).

On the other hand, it provided a significant impetus to evaluate whether it points to the result of the technology's diffusion at various levels through different routes (see Hall 1987; Tylecote 1975). This study also revealed the peculiarity of iron production technological apparatuses *vis-à-vis* the immediate neighboring regions of the study areas (e.g. Haaland *et al* and Todd's works in the Omo valley).

In the following, I will present the various technological variables (*cf. chapters III & V*) in iron smelting in order to provide insight into means of interpreting vestiges of metallurgy in archaeological context. For this, I also employed previous research outputs (e.g. Cline 1937; Haberland 1959; Haaland *et al* 2000, 2004a; Huntingford 1955; Lewis 1965; Todd 1985; Todd and Charles 1978a, 1978b) in different parts of Ethiopia despite their paucity in terms of details. Notwithstanding the difficulty of observing in the archaeological record, iron smelting in Wollega also provided an opportunity to understand the human cognitive behavior in their engagement with their environment as reflected through various ritual responses in the *chaîne opératoire* of production activities. However, the most important input of this research lies in the identification of elements of ethnoarchaeometallurgical evidence in the archaeological record.

7.3.1. THE SMELTING HOUSE

The mental map of iron production in Wollega begins with the construction of the smelting house, *godoo buufaa* (lit. blowing hut). One of the research sites in Wollega (i.e. Aira-Gulliso) presented a rich perspective towards our understanding of how smelters organized and conducted their work spaces. Among many African regions smelting took place either under the shade of a tree or under an open walled simple hut or in an open space (see Barndon 2004; Childs and Killick 1993; Haberland 1959; Haaland *et al* 2000; Herbert 1993; Schmidt 1997; Todd 1985).

Nevertheless, smelting in Wollega took place in a 'permanently' built house (Burka 2008; *cf. chapter V*). This in turn indicates that the smelting site has been relatively permanent. The Wollega ethnoarchaeometallurgical evidence provides a fresh model to archaeological investigation and interpretation of metallurgy sites. The 'permanent smelting house' view is consistent with the local tradition indicative of successive occupation by families of smelters (cf. chapter IV). Hence, a deeper time view of regional culture-history can be obtained from metallurgical sites.

On the other hand, the only archaeological evidence in northeast Africa, where we have smelting conducted inside a room or surrounded by wall, is Meroitic iron smelting (see Shinnie 1980, 1985; Tylecote 1975, 1980), believed to be a centrally controlled activity (see Haaland and Haaland 2007). The difference, however, lies in that the Wollega enclosure is a wooden wall, not a wall made of bricks, whose preservation in the archaeological record is unlikely, except for possible remains of postholes.

According to the ethnographic evidence, the smelting house provides a ground upon which the smelting schema is founded. It determines the location of the furnace, the blowers' position, and other spaces such as ore processing, clay mixing, charcoal stacking, feeding etc leaving their respective impressions inside the production areas. The height of the furnace from the ground is also concomitant with the height of the ceiling of the smelting house. The smelting house is a high ceiling domed shape (*cf. chapter V*). Who could have imagined, in the archaeological record, that a smelting house was structured in such a way without the insight from the ethnoarchaeological evidence (*see* Figure 71)? The distribution of activity areas inside the smelting house include the furnace structure, the clay fermentation pit, ore selection and crushing site, and charcoal selection site. It is also possible to find a burned soil signature between the furnace front aperture and the front door of the smelting house.

It is very important to note that the evidence from ethnoarchaeological reconstruction of the smelting house provides an insightful interpretive tool for the archaeological investigation of iron metallurgy sites in the region under study. The survey in Wollega (in both the northeast and southwest) has revealed a number of former smelting and forging sites in different parts of the study. Remains of former iron production sites were identified. Today, debris of tuyère remnants and slag mounds are exposed in different localities due to construction activities (cf. chapter VI). Future excavation of selected sites would enable to establish a deep time and wider space perspective of iron production in the area.

The ethnoarchaeometallurgical record in these two districts of Wollega has provided considerable valuable information on the history of iron metallurgy. Firstly, descendants of former smelters could establish the link through genealogical counts of members of smelter families. In addition, it has informed us about the change that took place immediately before smelting was abandoned in the last 40-50 years ago. Before the 1960s,

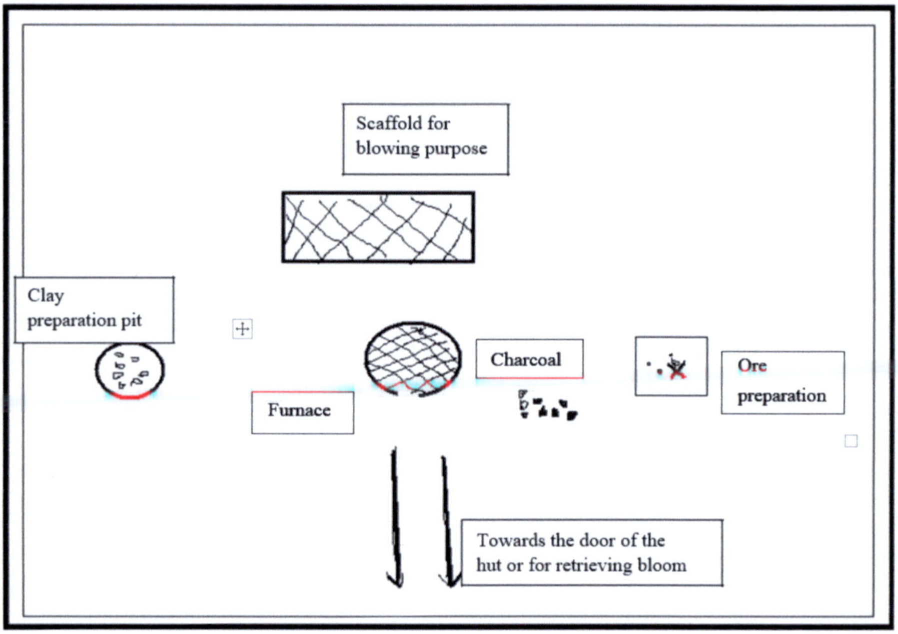

Figure 71 Sketch of space utilization and distribution of activity areas inside the smelting house in Wollega, Aira-Gulliso

former smelters were said to have dropped smelting activities. Former smiths took over the smelting activity that continued for sometime in the 1960s.

The other ethnoarchaeological evidence enlightening the past activities lies in its role to distinguish between the people and localities connected with iron smelting and smithing (see Burka 2006, 2008, 2009). Iron smelting reconstruction/ re-enactment in Wollega has revealed that there is clear distinction between iron smelters and forgers of the smelting products. This implies that smelting activity areas are separated from smithing areas. In other words, smelters and smiths occupied different occupational identities and hence separate hamlets (cf. *chapter II*). Accordingly, the habitation and activity areas of each identity are separate and discernible in the archaeological landscape. What does this signify archaeologically?

The Wollega model has demonstrated and become important for archaeometallurgical site interpretation in that the two activities have been conducted in different villages (hamlets). This was until the period when smelters decided to drop their former occupation probably after the Second World War and former clients of smelters adopted the activity in addition to smithing. In northeast Wollega, major concentration of ore-miners and smelters included Faallee and Waddessa Daanjee hamlets while smiths mainly lived and still live in Masoo and Somboo villages (*cf. chapter II*). These hamlets are separated from one another by creeks.

A similar distinction is also found in southwest Wollega. Whereas smelting activity areas concentrated in and around present day Aira town, the major smithing activity is located in the Abayyanii locality some 5-7 km away from the smelting and mining sites. The possible discovery of smelting remains inside the hamlets where smiths dominate today should cautiously be interpreted. This is because; it might have been part of the irregularities assumed to have taken place in Wollega in the post-WWII period, following the gradual dwindling of its activity in former smelting localities.

It is clear that the archaeology of metallurgy involves the recording and analysis of vestiges of human activities concerned with production. There is no problem with mining sites as opposed to what is related to smelting or smithing activity areas. It is at this critical level that evidence from ethnoarchaeology becomes valuable to create a clear picture of how humans, since the distant past, have structured their production activities. One challenge in the archaeology of metallurgy is the difficulty of distinguishing between smelting and smithing sites. This challenge originated from lack of ethnographic evidence to enable us to project a diachronic perspective. In addition, some ethnographic data where both activities are conducted by same individuals (e.g. see Haaland *et al* 2000, 2004a; Todd and Charles 1978b, Todd 1985 in the Omo Valley) blurred the picture and led to wrong conclusion that smelting and smithing might have been of the same origin or that both metallurgy sites could overlap.

The Wollega example is, therefore, an important model for us to understand the significance of separately addressing the two activities. According to our ethnoarchaeological evidence, there is no locality close to Aira town in the southwest or Waddeessaa Daanjee in the northeast wherein smithing took place in the past or where smiths inhabited interspersed with smelters. In other words, metallurgical vestiges in these localities belonged to results of various smelting activities owned by different families. It is still necessary to identify how the smelting house (cf. chapter V; Figure 71) activity areas enable us reconstruct human behavior in the archaeological record.

As previously noted, the smelting house is a circular one. One of the smelting houses of the reconstruction project in southwest Wollega (Burka 2008) faces towards the east while the second one faces in the opposite direction. Inside the house, the front side of the furnace faces towards the front door of the house (*see the two arrows in fig.71.*). In the course of the ethnoarchaeological research, it became clear that the direction of the door, hence the furnace, is determined by the topography (or the slope) of the ground upon which the site is established. A similar decision was observed for the smelting site in northeast Wollega. The location of the smelting house (and furnace) is to ease the work of the blowers and flow of wind. More important in this arrangement is to facilitate blowing activity. According to informants, it is easy to blow 'downwards.'

Thus, the design of the house determines the position of the furnace, the direction of its 'front' side and the location of blowing place. Where is the furnace situated in the house and why (see Burka 2008)? How is this useful to provide clues in the attempt to understand one stage of operational sequences in iron metallurgy? Nevertheless, before these, there is a question to be addressed regarding the decision of the smelters to conducting iron smelting inside a house. It raises another issue whether or not iron smelting used to be conducted inside the village in the past.

Given the location of the smelting sites (particularly in southwest Wollega) within or close to habitation areas (from today's observation), it can be posited that iron smelting has not been regarded as polluting by the villagers or the smelters themselves. However, it needs a wide range of archaeological investigation to establish ancient settlement patterns of the areas in order to determine whether smelting used to be conducted inside the villages.

The smelting house is about 20 meters tall at its apex, primarily supported by a single pillar, which is gradually cut parallel to the wall top across which a wooden beam is laid. The pillar was erected in the middle of the room that divides it into equal parts. In the traditional countryside house, the place where this temporary pillar is erected is later on used as the 'common' fireplace. It is the fireplace where the head of the household, guests and children gather and where informal education is passed to the children in the form of stories, riddles and the like.

The family fireplace is built in bowl shape with beaten clay. Three stone supporters (*gomjii*) are erected inside the fireplace at equidistance from one another. One of the stone supporters is situated perpendicular to the main door and is named as the stone of '*Maaramii*' (*gomjii maaramii*) (*cf. chapter VI*). '*Maaramii*' is a fertility spirit and rites related to fertility such as post-child delivery initiations are conducted close to *gomjii maaramii*.

Although the fireplace is a common meeting ground for the family and strangers, on other grounds, it symbolizes and represents the activity areas that ensure women's fertility. Despite its being a common meeting ground, there are circumstances under which individuals are barred from sitting near the fire with the rest of the family. This includes individuals who are inflicted with sexually transmitted diseases (STD). According to the belief, fire or heat creates a favorable ground for the transmission of the disease from the victim. Fire is an agent of transmission. It transforms things from one state to another irreversibly e.g. iron ore to bloom, natural to cultural. Like iron, fire is ambivalently understood. Fire is both a symbol of life and death. Fire symbolizes human fertility. It is also a powerful force of destruction.

In the iron smelting setting, the fireplace in the center of the house is where the furnace is laid. As the fertility stone supporter in the fireplace faces towards the front door of the house, the anthropomorphic furnace is situated in that its front part is facing the door. It is clear that iron smelting is a male domain. There is no practical involvement of the women in iron production activities. Women provide food for male workers during the smelting. However, metaphorically women[77] occupy a central place in the iron-smelting schema. The furnace (e.g. southwestern Wollega) with all its body parts symbolizes a woman and the responsibility of the smelters is to invoke her fertility for attaining successful production.

77 Although women were not direct participants in iron metallurgy, there is the story that relates them to the activity-this includes the story of the granary maker, the potter (chapter VI) and the like. In addition, in northeast Wollega, a story attests that the wives of the chief among the Malkee clan did own their own respective smelting houses. According to one of the informants, his mother owned two smelting houses, which were operated by poor members of her family, one of whom was her brother who served as a master smelter.

Archaeologically speaking, the setting in the house might clearly be distinguishable. The location of the furnace in the house, the direction of the front side, and the position of the blowers etc can indirectly be reconstructed. Whenever it is distinguished, however, the thought of the society and the philosophy of their space use in the house should need serious consideration. Another point I would like to emphasis is the likelihood that smelting which took place in the house signifies. It likely signifies that iron smelting was not, at least once in the past, a marginal socioeconomic activity. This would suggest that the story of smelters belonging to the upper echelon of the society is well founded (see Burka 2006, 2008, 2009).

The inside house activity area organization also provides a number of interpretative models for archaeometallurgical inquiry of activity sites. For instance, in one of the smelting houses a circular pit (*cf. chapter V*) of over half a meter depth in which clay soil was dampened and mixed with tempering materials, was dug into the ground. Since its major importance is during furnace building, in the subsequent period, the pit might end up to serve as a dumping place for discards from smelting including ash, charcoal fragments, slag etc. This possibly gives a wrong impression that the pit was meant for a furnace base.

Similarly, the underground furnace of northeast Wollega could challenge archaeological interpretation. The challenge is that it can be regarded either as a pit upon which a furnace superstructure was built or the pit part of a design into which slag was tapped. As a result, a number of suggestions of the size of its superstructure are possibly made.

7.3.2. TECHNICAL CERAMICS

The technical ceramics are mainly furnaces and tuyères. The identification of these iron production apparatuses is of paramount significance for understanding the history of metallurgy of the region. Furnaces pose several opportunities and challenges for the study of a history of African metallurgy. On the one hand, the apparatus has been regarded as a mechanism to understand the spread and development of the knowledge in space and time. Protagonists of outside origin (cf. *Chapter I*) have espoused this in particular. Conversely, the so-called 'African bloomery iron smelting' (derived from a coagulated mass called bloom; see Tylecote 1980; Haaland 2007) and its homogeneity was sought to explain the achievement of African societies in this regard (for critics see Mapunda 2000; Schmidt 1997; Schmidt and Mapunda 1997; Schmidt 2001).

The diversity in African furnaces has challenged the hypothesis that the technology throughout the continent had similar or single origin. One challenge, for instance, to the Meroitic 'Birmingham' hypothesis (Sayce 1912) was the lack of archaeological evidence in different directions (Mapunda 1997; Trigger 1969) outside the aforementioned site. Different types of technological approaches seem to have been developed in various parts of Africa. In light of this, the ethnoarchaeology of iron production in Wollega, and previous works in the neighboring regions including Oromo inhabited areas, reveals that the model to interpret culture-history of a society through the distribution of furnaces is untenable.

On the other hand, the 'migrationist' approach (see Sinclair *et al* 1993) that many scholars employed to explicate the means of cultural diffusion across the continent lacks plausibility. Does population migration or movement explain the technological history of iron production at least in Wollega and its immediate neighbors?

One major thesis[78] regarding population movements in Ethiopia is the so-called Oromo population migration of the 16th century from the highlands of Bale in southeastern Ethiopia to different parts of the country (see Gidada 2001; Hassen 1990; Megerssa 1994; Ta'a 1980; Zewde 1991). As elsewhere in Africa, some scholars have developed an argument in favor of historical processes including population movements to explain the origin of marginalized artisans and hunters in different parts of the Ethiopian region (see Freeman 2001; Pankhurst 1999, 2001). The Ethiopian 'historical process' model espoused by Freeman and Pankhurst in juxtaposition with other factors is short of addressing the issue through material or technological means.

While the discourse that rules out the caste model (Pankhurst 1999, 2001) for the emergence of marginalized groups on the basis of lack of common origins of the groups is plausible, it did not find out whether the technology adopted by each minority group in each society can be comparable and traceable across wider spaces. Supposing that the Oromo moved in the sixteenth century from southeastern Ethiopia all the way north and westwards to settle in various regions of present Ethiopia, a number of assumptions can be developed. This also takes into consideration the possibility that the Oromo predecessors had already adopted and developed iron production technology in their respective niches. Furthermore, it is in addition to the significance of the consideration that the Oromo have used a formidable cavalry force to defeat highland and lowland kingdoms and principalities equipped with the then 'modern' European firearms.

However, the consideration should also include the argument that the Oromo were armed with 'horn spear points'[79] (see Haberland 1963; Todd 1985; Tylecote 1975 and for critics Burka 2006) during their 'en-masse' movement across the Ethiopian region. Thus, if the Oromo were instrumental in homogenization (see Hassen 1990; Zewde 1991) of the Ethiopian population, then one of the testing grounds would be found in the technology of iron smelting to be discernible across the board.

However, much to the disappointment of writers who espouse the 'migration' model in the spread of various cultural attributes including technology, available data in many Ethiopian region and the Horn (particularly in Ethiopia) underlie that there existed many instances of unrelated technological apparatuses among closely connected societies. Four examples

78 The antithesis (oral history collected by Chernet Waqweyya) of the 16th century Oromo movement and settlement in the highlands of today's Ethiopia such as in Wollega is the genealogical count of over 600 years ritual and political sites established in this regions. An example of this is the descendants of Jaawwii, who is said to have founded the Bulluq Gadaa center in Horroo highlands of northeast Wollega. By the early 1970s, the last Gadaa group (e.g. Galataa Abbaa Namaa) is said to have received the 68th cycle of the system, which means 68*8=544 years old since the first Gadaa at Bulluq.

79 There is a distinction among age groups in the type of spear points and sizes used at least in the Oromo of Wollega. What is in particular important is the fact that real spears points (i.e. iron heads) were not allowed to be held by youngsters who had not developed experience in using them. Instead, wooden spear points or horn points of bucks are used. The observation of this practice may have created the wrong impression that it happened due to lack of the use of metal tools.

inside Oromo inhabited regions of central and southwestern Ethiopia provided empirical evidence that point to clear divergence in the technology of iron smelting particularly the furnaces- the most important technological variable and the powerhouse of traditional iron metallurgy (*cf. chapter III*). These are in central Ethiopia, among the Shoa Oromo (Cline 1937), in the Kingdom of Abba Jifar of Jimma in southwestern Ethiopia (Huntingford 1955; Lewis 1965), in northeastern Wollega (Burka 2006) and in southwestern Wollega (Burka 2008).

The furnace in Shoa is markedly different from the rest particularly in terms of operational or charging mechanisms. The furnace is primarily a natural draught one never recorded in Ethiopian region and the Horn. This furnace, which was over two meters tall was constructed at mining sites and used only for a single smelt (*cf. chapter I&V*). There are no details as to how much of the part of the furnace was destroyed after each smelt and whether it was due to practical or symbol reasons. Traditions and historical sources claim that there was once a movement of people westwards (as well as eastwards) from this center (e.g. see Gidada 2001; tradition of lineages).This could have opened various ways for the spread of the knowledge and its appearance west of Shoa in Wollega, for instance.

In terms of geographical proximity, northeastern Wollega could have been a right choice to find similar technology of iron smelting as in Shoa. On the contrary, however, it provided a rather distinct iron smelting technology never recorded in Ethiopian region and the Horn (Burka 2006). The other closest region to Shoa in the southwest is former Kingdom of Jimma Abba Jifar south of the Gibe river (see Hassen 1990; Huntingford 1955; Lewis 1965). Similarly, it could have been possible for technological exchange or transfer through the movement of people and ideas from Shoa to the Jimma area, where the activity has been documented.

In contrast, the iron-smelting furnace in Jimma areas (in the Gibe river valley) was a forced draught apparatus. The only similarity with that of the Shoa example (Cline 1937), is that the furnace was built at mines and demolished after every smelt (see Huntingford 1955; Lewis 1965). Nevertheless, here also, there is a gap of information on the level of destruction. Given the fact that the Jimma Abba Jifar iron metallurgy was an activity controlled by the king's representatives and the kingdom constantly sought the products, a regular demolishing of a furnace is uneconomical. The product was regularly taken to the smiths who lived in Jiren, the capital of the kingdom.

On the other hand, according to the evidence from Jimma, smelting was not only conducted close to the mines but smelters and smiths had separate habitations. This implies that any archaeological investigation of the area around the palace in the form of metallurgical remains provides data not on the components of smelting activity but only on the smithing. The challenge is that it might lead into a wrong interpretation that raw iron was imported from elsewhere outside of the kingdom.

There are two problems for archaeological investigation and interpretation of such production sites of iron metallurgy (i.e. in Shoa and Jimma). First, since furnaces are always 'destroyed' after each smelt, it implies that there is difficulty of recovering them in the landscape. As a result, it would not possible to conceptualize the technological and cultural aspects surrounding the furnace. Nevertheless, if vestiges of a furnace could be found *in-situ*, it would not tell us more than the date when the last smelting was conducted immediately before the site or activity was abandoned.

The other problem relates to the use of furnace fragments for the subsequent activity. As many experiences such as in pottery works indicate, there could be a possibility that furnaces regularly demolished might have been in various ways reused in the subsequent smelting activities. In view of lack of precision of carbon dating of charcoal remains (e.g. due to contamination, old wood), tempering organic materials could yield alternative means of chronometric dates of technical ceramics. However, in light of the possible reuse of the materials, archaeological interpretation of such sites could be as complicated as carbon dating from charcoal remains.

As I have pointed above, the cultural meaning surrounding iron smelting in these two regions of central and southwestern Ethiopia, south of the Gibe river could have totally been lost with 'destruction' of the furnaces. Ethnographic sources in Africa have shown furnaces have contained and been used as medium for transmitting ideas and thoughts of the society. This partly takes place through decoration and body marks of the furnaces, etc. as the Wollega ethnoarchaeological material has revealed (Burka 2006, 2008).

To the west, the first close iron production site to central Ethiopia is located in northeast Wollega of Abee-Dongoroo district (Burka 2006, 2008, 2009, 2010). Ethnoarchaeological re-enactment of iron smelting practice before it stopped in the 1960s has provided insights to understand the distribution pattern of production activities namely iron ore mining, clay extraction, charcoal making, smelting and smithing sites. In particular, the iron-mining site is closer to the village that monopolized smelting than where smiths concentrated.

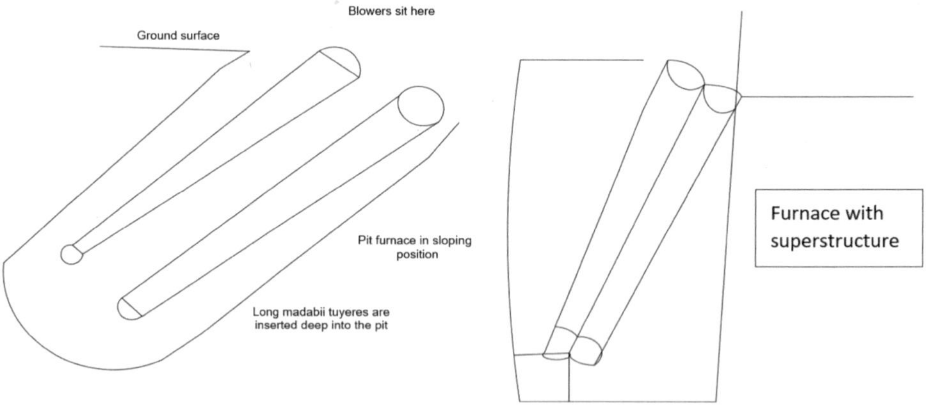

Figure 72 a cross-section of underground furnace of Abee-Dongoroo, northeast Wollega (left) and upper-ground furnace of Aira-Gulliso, southwest Wollega (right). Note that the two clay tuyères are inserted deep inside the furnace.

According to local tradition, smelters and smiths occupied separate occupational and social identities until late in the 1950s for a short period of time when some smiths also added smelting to their original skill. This is attributed to changes in relationships during the Italian occupation of 1935-1941 between smelters and the chiefly class in the Malkee clan to which the smelters belonged. Consequently, it appears that after the 1940s, the intensity of smelting activity has diminished and gradually taken over by the smiths/*tumtuu* with intention to supplement the demands for the small supply of scrap iron.

In complete contrast to the type in Shoa in the east, the Abee-Dongoroo smelting was conducted in an underground furnace (Burka 2006; Figure 72 left; cf. *chapter V*). This shows that the two smelting sites did not have any common technological origin. Conversely, however, it is possible to posit that the northeastern Wollega smelting apparatus belongs to a 'primitive' level of technological achievement leading to a more 'sophisticated' technology as the one in southwest of the region (Figure 69 right). The only close similarity with the Abee-Dongoroo furnace is the so-called Pre-Roman iron smelting technological apparatus at Meroe (Shinnie 1980, 1985; Todd 1985; Tylecote 1975, 1980).

However, the problem with the Meroitic bowl-like structure is the absence of evidence clearly distinguishing whether it belonged to smelting or smithing. The discussion on the Meroetic iron production (see Shinnie 1980; Tylecote 1980) concentrated on the search for similar identities elsewhere than attempting to focus on local variants. There is lack of clear evidence on the pre-first century AD iron-smelting furnace at Meroe. Nevertheless, the geographical location of Abee-Dongoroo in the Blue Nile river system of the Dhidheessa valley tempts to point to the direction of origin of the knowledge. This conclusion partly emanated from lack of technological similarity towards the east from where the Oromo groups might have 'moved' westwards. These possibilities further tempt to envision two scenarios for the origin of the northeast Wollegan underground iron smelting technology.

First, it might have been possible that before the Roman type of slag-tapping furnace appeared, areas all the way along the Nile regions in the Nile river system could have developed their own smelting technology, which employed an underground furnace. Secondly and related to the above is the possibility that the Abee-Dongoroo furnace could be the result of local experimentation through previous pyrotechnological knowledge. The similarity between the smelting furnace and the local mining pit as well as the local smiths' myth of origin related to the potter's kiln incident (*cf. chapter V*) leading to its discovery tempts us to prefer this suggestion.

In the absence of sophisticated metallurgical analysis of sites and finds, underground furnaces as in Abee-Dongoro could pose a serious challenge for archaeological interpretation (see Burka 2006). Such a furnace could give the impression that the pit might have been an underground extension to tap the slag from the furnace. As the furnace superstructures are prone to destruction due to various factors including wet climate during the rainy season as in Wollega, such a conclusion is not unlikely. Regarding space utilization around the furnace, there is much similarity with the one in southwestern Wollega. Blowers sit on high ground to pump downwards. Since the furnace is in the ground, there was no need to construct a bed for bellows workers. Therefore, no postholes close to the pit furnace could be identified as opposed to its counterpart in Aira-Gulliso. However, the slope of the ground is maintained in that blowing is held from up ground downwards giving the blowers the advantage of using their full energy for easily puffing the air into the furnace.

Southwestwards from Abee-Dongoroo iron smelting district and over 500 km away from Shoa, is located the Aira-Gullisoo iron metallurgy. Two issues are available for comparison between the two technologies in Wollega. Historical sources on the nature of regional population dynamics in Wollega since the 16th century AD (see Gidadaa 2001; Ta'a 1980) claim an east-west orientation of the movements and settlements. Local names of places and clans in east and west Wollega seem to attest to the presence of connection between the groups in both regions. Oral traditions also support the story. For instance, Malkee clan members in Aira-Gullisoo district are said to have migrated from their ancestral homeland in the northeast Wollega such as Abee-Dongoroo area. They are offshoots of the northeastern Wollega inhabitants. This approach could have served to trace cultural traits throughout the region. Accordingly, the distribution is from northeast towards the southwest originating in the former. This in turn implies that the latter is younger than the former.

However, the presence of two unrelated smelting furnaces in both districts disrupts such a schema for the cultural distribution in Wollega. Even then, the use by smelters and smiths in both regions of the same types of tuyères components and bellows provides a ground for arguing in favor of connections between the people in charge of the knowledge. African archaeometallurgists argue that pit or bowl furnaces represent elementary or 'primitive' form of iron smelting history (see Cline 1937; Haaland 1985; Kense 1985; Shinnie 1980; Tylecote 1975; 1980) as opposed to furnaces with superstructure. This implies that the furnace in northeast Wollega is older in date than the much more 'sophisticated' one in the southwest parts of the region. Before I suggest the possibility of their origin, I shall discuss in the following section a little about the peculiarity of the Aira-Gulliso area furnace (Figure 72 right) based on the ethnographic observation I made during the reconstruction of the smelting activity (Burka 2008).

Our knowledge of the technological aspect of metallurgy in southwestern Wollega is much deeper than in the northeast (cf. *chapter V*). In this part of Wollega, former participants with vivid memory of former smelting practices are still alive (Burka 2008) unlike in the northeast where only two brothers of close to 90 years[80] old could provide information that they remembered (Burka 2006).

In Aira-Gulliso, it was much easier to locate former smelting sites due to various factors. On the one hand, iron smelting practices have continued very late in the 1960s because of which eye witnesses are still available. In addition, many former metallurgical sites are being exposed due to human and natural factors. Furthermore, unlike the dense forest and wet environment in the northeast, sparse vegetation in the region has helped surface investigation of sites that could corroborate traditions.

80 The two brothers in Abee-Dongoro were the sole providers of the information on former traditional iron metallurgy that led to its reenactment practices in the summer and autumn of 2005. They have also served as key living testimonies to distinguish between the position of former iron smelters and smiths, and their habitation sites. As I have stated previously, one of the two brothers is now dead.

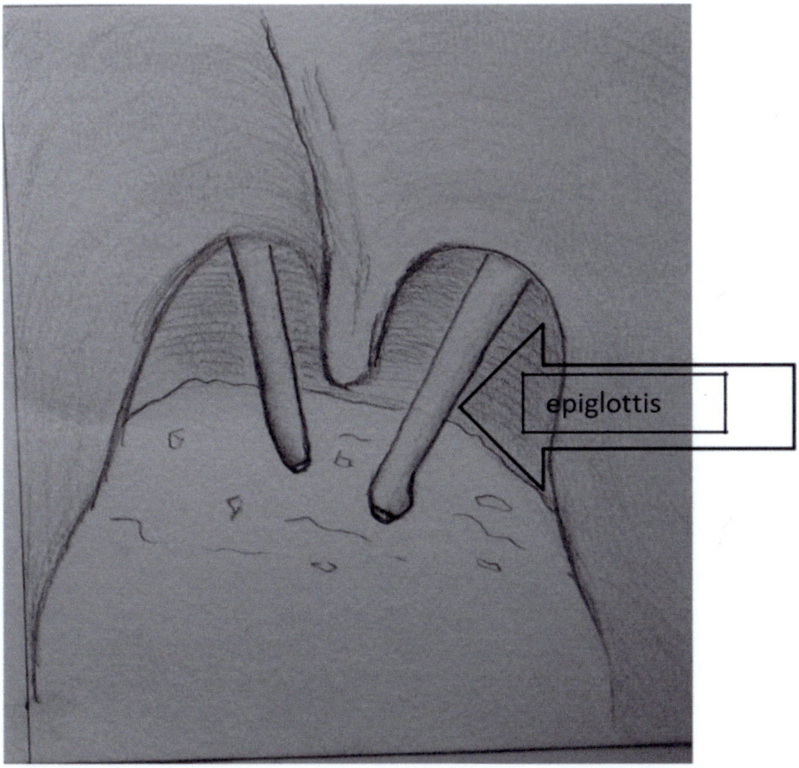

Figure 73 sketch of Aira-Gulliso's upper part of furnace observed through the 'mouth'. Note the alignment of the two tuyère arrangements leaving the 'epiglottises' in the middle.

The furnace is built in the centre of the smelting house (Burka 2008) in a circular structure. Furnace building begins about 50 cm underground in a narrowly dug channel for erecting the wooden wall leaving a front door. This design is similar to the structure of traditional thatched smelting house. The channel is filled with clay soil to tauten the wall and simultaneously the wall is built upwards and tied up with creepers until the desirable height is attained (*cf. chapter V*). Before the inside wall is plastered with wet clay, which is tempered particularly with chaff of cereals such as *teff* (teff *eragrostis*), finger millet (Eleusine *coracana*) etc the wooden wall is burned out leaving a charred and channeled surface to be filled with the plaster. The wall is made from young flexible wood.

In addition, the straw tempers are useful to determine not only the changes and continuities in the techniques of furnace building but to provide also alternative means of determining the age of the site. According to the ethnographic evidence, two circular structures could be formed (and would leave postholes) in the setting where smelting has once been conducted. The wider circle represents the house structure and the narrower and central one is for the furnace structure. This is in addition to another circular pit for clay processing in the operational sequences leading to furnace building.

The ethnoarchaeology of iron smelting reconstruction in southwest Wollega led to the building of the furnace with various body decorations and markings. These include the breast-like extension, ears, epiglottis and glottis, and a rectangular bottom aperture as well as the chimney (mouth) and two circular holes at shoulder length. These would serve as tools for tracing the sequence across time. During smelting, a sacrifice including incense burning in front of the bottom aperture as well anointing the furnace body with blood should be considered during the examination of archaeological finds of metallurgical debris.

On the other hand, the most conspicuous archaeological finds in traditional metallurgical sites are slag fragments (see Chirikure 2005; Miller 2002). It was such remains in Meroe, for instance, that led Sayce (1912) to declare it as the 'Birmingham of Africa.' It is particularly because slag is the indestructible element of metallurgical remains (see Haaland 1985; Haaland and Haaland 2007; Shinnie 1985; Todd 1985; Tylecote 1975). The slag separation and its accumulation provide a clue towards the smelters' technical capability and the quality of the ore.

In the Wollega iron-smelting context, the separation of the slag takes place at two levels (*cf. chapter III and V*). The first stage is in front of the smelting house or at the smelting setting after the bloom is dragged out of the furnace. After this preliminary refinement, the unsorted mass of bloom is taken to the smiths where the slag is finally beaten off.

Archaeometallurgical analysis of slag in different localities in Africa made possible the identification of slag formed in either smithing or smelting context. Its analysis also helps to determine the provenance of the ore smelters used (see Todd and Charles 1978b). In addition, it enables us to analyze the achievement of the smelter in using a mechanism to separate the impurities during smelting. Slag can be formed in the form of cake (see Chirikure 2005; Haaland 1985, 2004; Todd 1985; Todd and Charles 1987b) if the furnace is built on a pit or of flow structure and is formed when the channel is set to tap it. It is from this aspect that the division of furnaces into what is known as slag tapping or non-slag tapping was formed (*cf. chapter I and III*).

Another technical ceramic conspicuous in the archaeological record of metallurgy is the clay tuyères (see Figure 74). The tuyère remains are informative of technological aspects of metallurgy in various ways. The mechanisms include the components of tempers for their making, and the level of penetration into the furnace inferred from slagging (see Chirikure 2005; Miller 2002) formed (at the) on the distal side of the tuyère remains. The examination of the latter aspect helps to determine the level of temperature attained during the smelting and the skill of the smelter in controlling the heat in the furnace. In addition, charcoal fragments trapped at the tuyère ends with the slag are useful to reconstruct the environment contemporaneous to iron smelting time as well as for carbon dating it. Slag and tuyère remains in archaeological context confirm the presence of local metallurgy more than simply a hoard of metal tools. Archaeometallurgical analysis of metallurgical remains and metal tools can provide the means by which the products were attained (*cf. chapter III*).

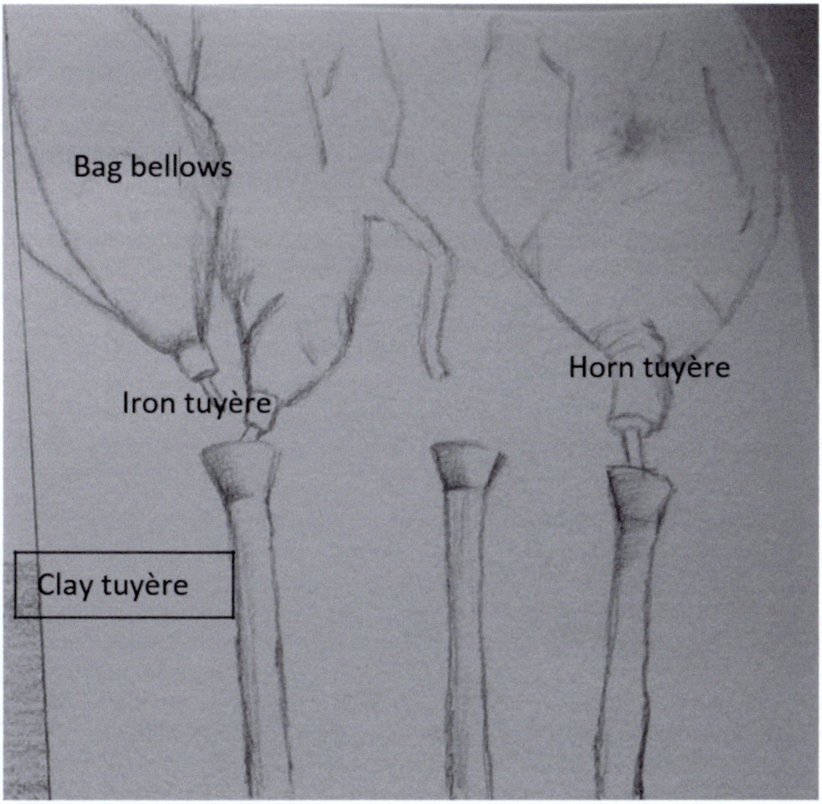

Figure 74 sketches of tuyères and bag bellows component. Note the tuyères consist of clay (madabii), iron (kololii), horn (buudaa, mundoo).

Archaeological survey in the two research areas of Wollega under study has yielded that iron metallurgy has once been an important aspect of the economic activities of the regions (cf. chapter IV). In particular, ore mining scars left on the landscape in both northeast and southwest Wollega showed that there was an intensive exploitation of the environment and that iron products were largely sought by the population of the region or beyond. It confirmed that Wollega served as an origin from which iron was traded across western and northwestern Ethiopia and into the Sudan as far time back as at least in the 19th century (see Ahmad 1999; Triulzi 1973, 1975, 1981; Waqweyya 1988).

In addition to ore mining evidence, many former sites associated with iron metallurgy were identified. According to the survey results, former smelting sites are located close to streams but not necessarily to the ore mining sites. This is particularly evident in Aira-Gulliso district where many former smelting sites are concentrated. Today, human and natural agents have resulted in the exposure of mainly tuyère and slag fragments that formed debris in Aira town (cf. *Chapter IV*). Mounds formed constitute slag and tuyères fragments. Aira town former smelting sites are located hours of journey from mining sites ranging between 1:30 hour from Wiinsaa Qachoo to over 2 hours distance to Warra Qaalluu and Caaggo and about 4 hours from Dimmaa one way trip across foot tracks.

Nevertheless, in the northeast, the major smelting hamlet (Waddeessa Danjee) is located less than an hour from the ore-mining locality at Faallee. This does not rule out, however, the possibility that iron ore has been traded outside of Abee-Dongoroo (*cf. chapter VI* on the story of the granary maker woman) in various parts of northeast Wollega or in the southwest.

Unlike in Abee-Dongoroo where mining was limited to a single site, the smelters of Aira-Gulliso district exploited four major mining sites. From the point of the location of the mining sites *vis-à-vis* the smelting areas, the Wiinsaa Qachoo mining site looks the most favorable choice (see Wilson 2007 on the constraint of terrain difficulty). In addition, laboratory analysis of the ores from Wiinsaa Qachoo, Warra Qaalluu, Caaggoo, and Dimmaa (Hamrla 1966; *cf. chapter II and III*) showed convincingly that both sites provide similar quality of iron ore. According to Hamrla (1966:447), the chip sample analysis of ore from Yubdoo (Dimmaa), which is very far and (K)Qachoo which is relatively close provided 70.9% and 70.8% of iron respectively.

Therefore, the *saakuree* ore story, (*cf. chapter V*) attributed to Wiinsaa Qachoo ore to prefer to travel for hours to bring it from Yubdo locality, does not seem to have any technological grounds. On the other hand, I have made a laboratory analysis of ore samples from different localities to test their melting point at Addis Ababa University, Department of Chemical Engineering (analysis by Biruk Yohannes, Laboratory Technical Assistant). This gave a rather different picture of the two mining sites from whence smelters obtained ore. Whereas Yubdo's ore could melt under a temperature in the range of 1335-1350°C, the ore from Wiinsaa Qachoo did not melt until the temperature reached 1500°C. This might have been a reason why smelters avoided the latter in favor of the former site. However, future detailed laboratory analysis of carefully collected samples in former smelting sites *vis-à-vis* the iron ore in both sites would provide whether smelters' decision to depend on one mining site rather than the other was due to cultural preferences or technological necessity.

Furthermore, in addition to locating metallurgical sites, the ethnoarchaeology of iron smelting in Wollega could be used as a model for the reconstruction of the social organization of various occupational identities in the past. Despite changes due partly to migration from rural areas into nearby towns and partly due to the Derg's resettlement projects, there still exists clear evidence indicating the influence of economic activities for concentrations of habitation sites in separate localities. According to ethnoarchaeological evidence in both Wollega study areas, the smiths and smelters inhabited different hamlets. This implies that the vestiges of metallurgical activities in the hamlets of the smiths may most probably correspond to the result of smithing (of locally smelted bloom or imported scrap iron) while the remains where descendants of former smelters concentrated is attributed to smelting activities.

Wollega's technological peculiarity and variability is also found in the three types of tuyères made of clay, iron, and horn. According to scholars, bag bellows are common in northeast Africa. Bag bellows are also considered the highest level of achievement in the development of bellows(see Cline 1937; Haaland 1980, 1985, 2004a, 2004b, Haaland *et al* 2000; Kense 1985; Miller and Van der Merwe 1994; Pleiner 1980, 2000; Shinnie 1985; Todd 1985; Todd and Charles 1978a, 1978b; Tylecote 1975, 1980; Wainwright 1942a, 1942c; Van der Merwe 1980). However, evidence on arrangements such as in Wollega that employed a succession of three types of tuyères (Burka 2006, 2008, 2009) is lacking. Bag (skin) bellows are likely the least possible survivors in the archaeological record.

Likewise, tuyères made up of horn are prone to decomposition and destruction. The fate of iron tuyères might not be so different from the horn ones given the wet and humid climate that would destroy it because of corrosion. The most likely survivors in the archaeological record are fired clay tuyère remnants. This explains why these are among the most conspicuous metallurgical remains. Experience from ethnoarchaeological study in Abee-Dongoroo showed that tuyères left over from the smelting were highly sought by smiths. In addition to affecting its analysis, this explains why they are sparsely distributed in the archaeological landscape. The chance of their reuse is higher where smelters and smiths live very close to one another or same persons performed both activities.

The high concentration of tuyères in former smelting villages in Aira-Gulliso might be due to the distant location of the hamlet of smiths. Such factors could blur the way we look at and interpret archaeological site formation processes in metal production sites. The ethnoarchaeological data in Wollega on the types of tuyères could be used as a model in that the absence of tuyères other than those made of clay in the archaeological record should not be taken to mean that the smelters or smiths employed only one type of tuyères.

On the other hand, the bellows are made from goat or buckskin (Burka 2006, 2008; Huntingford 1955; Lewis 1965). Such bellows are said to have originated from the Middle East and are said to represent a higher level of development of the apparatus compared to the pot or wooden bowls (Cline 1937; Kense 1985; Haaland 1985; Shinnie 1985; Tylecote 1975, 1980; Van der Merwe 1980). The ethnoarchaeological evidence from Wollega that showed that smiths have improvised car tubes to make bellows out of them, should be viewed in light of long period of experimentation and innovation on the various types of bellows before the bag variants were adopted. The continuation of pot bellows in the Omo valley (Haaland 2004a; Haaland *et al* 2000, 2004b; Todd 1985) of southwestern Ethiopia could be attributed to the region's remoteness to be affected by interactions or the artisans' conservative behavior to continue to maintain it.

7.4. EVIDENCE FROM WOLLEGA: SOCIO-CULTURAL

In the previous section of this chapter, the major focus was on the identification of variability in terms of technology of iron smelting in various parts of the Oromo inhabited regions namely Shoa, Jimma and Wollega *vis-à-vis* their Omotic-speaking neighbors in the Omo river system. The result from ethnoarchaeometallurgical records, survey of former smelting sites (for Wollega) and literary sources (for Shoa and Jimma) revealed that the evidence from four sites provided unrelated technological variables in iron smelting operational sequences. A striking challenge to the 'single' origin assumption and hence the search for 'homogeneity' in the technology of iron smelting was found among closely related lineages whose history is connected with iron metallurgy in two districts of Wollega.

However, crosscutting this technological variability, we find similar socio-cultural aspects surrounding iron metallurgy in these two areas of Wollega. Many of the socio-cultural aspects we identified in the Wollega context seem to be of panhuman means of elaborating ideas being spun around iron metallurgy and its workers elsewhere (see Burka 2008; Collett 1993; Haaland 2007; Haaland and Haaland 2007). The challenge is how to explain this archaeologically. It is also difficult to explain it in terms of migration and settlement of similar people or diffusion of ideas. This is already identified in the section on technological aspects of iron smelting that challenged the issue of homogeneity.

In light of the above issues, the challenge remains pervasive since such crosscutting similarities of socio-cultural aspects of iron metallurgy is found among geographically unrelated communities in many parts of the world (see Barndon 2004; Burka 2006; Chirikure 2005, 2007; Collett 1993; Haaland 1980, 1985, 2004a, 2004b, 2007; Haaland and Haaland 2007; Haaland *et al* 2000, 2002, 2004). This should not, however, be taken to rule out the possibility that contacts through movements of people, ideas and objects could have created outside impressions on the local societies. Thus, the local communities in the end contextualize outside elements to integrate them into their local variants (see for example Manzo 2009). In other words, they could adapt outside elements of culture into their locally diverse environmental requirements (see Chirikure 2005; Mapunda 2000; Schmidt and Mapunda 1997; Schmidt 1997).

In the following discussion, I shall concentrate on objects involved in iron production and metal tools and weapons in forging various cultural meanings in their different settings (e.g. ritual vs. secular). First, through technological variables in iron smelting (e.g. bellows, tuyères and furnaces), I shall try to explain the socio-cultural and economic basis of iron production in the society. Then, I will discuss some metal tools and weapons to interpret the intriguing social status of the artisans. This is derived from how the critical role they play among the people they lived could lend to the way the same society being served has perceived them ambiguously. The changing position (positive and negative) attributed to iron workers is derived from the ambiguous positions occupied by their products as tools of production and weapons of destruction (see Burka 2009; Haaland 2007; Haaland and Haaland 2007).

The bag bellows (*qalqala buufaa*), tuyères namely clay tuyère (*madabii* or *qunxurroo*), iron tuyère (*kololii*), horn tuyère (*buudaa, mundoo*) and furnace (*gumbii, boolla buufaa*) provide the bases of meaning particularly the material-cultural one intended to be spun around for elaboration of ideas through objects of production or use. The cultural meanings are discerned at various levels of the *chaîne opératoire* including smelting, smithing and use. Metaphors attached to the objects enable new meanings to be created and transmitted. The new meanings embodying as important function as the original names.[81] What is in the names? The impressions one can find in the names of objects of iron production is one of

81 For instance, *Qalqala* literally means sack, traditionally made up of skin or hide for containing cereals. In the context of smelting and smithing it refers to bellows. However, it has a third meaning, which is often hidden and appeared intentionally to connote a different meaning than its ordinary attributes. It is part of a male's reproductive organ, a scrotum. Thus, the bag bellows become part of the procreative process during iron smelting. This meaning is very clear from jokes being exchanged between the blowers and spectators who are waiting their turn.

production either human or agricultural reflected in tools and one of destruction brought about by weapons made through these processes. In the iron smelting setting, objects such as bellows, tuyères and furnaces provide cultural meanings as well as technical ones.

In terms of physical remains, it is not highly likely to find the bag bellows (*qalqala buufaa*) made of skin in archaeological record. However, the lessons from ethnographic record are apparently essential to find how the thoughts of the people are structured in production contexts such as iron smelting (cf. *chapter IV*). *Qalqala buufaa* is an Oromo phrase to represent an English equivalent of 'blowing sack'. Since sack is for containing things, in this case, it apparently refers to the air that the bellows contain. Beyond this ordinary meaning is the social attributes of sack in the context of parts of human body. *Qalqala* is an adjective term used to refer to man's scrotum. Metaphorically, the bag bellows represent the scrotum of the blowers who pump air into the furnace, which is further facilitated through the two parts of the penis (the *madabii* and *kololii* tuyères). Iron smelting in this procedure becomes closely connected with sexual activities. The sound created by the bellows is figuratively related to deep breath during sexual intercourse.

On the other hand, bag bellows are made from goatskin. There is a cultural reason for preference of goatskin to sheep skin. Goat is regarded as a *'property of the devil and cunning'* animal while sheep is regarded as *'innocent and dumb'*. The goat is needed for sacrifice to appease problems caused by some malicious forces (e.g. the devil). The sheep is in turn a sacrificial animal needed to maintain peace between people and their creator (e.g. *Waaqa*). There is a belief that goats kept by smiths are of high medicinal/conciliatory values. The goat is also regarded as a sexually potent animal like the smiths who are regarded as sexually 'hot' and 'dangerously' polluting (see Burka 2006, 2009).

The smith himself is a medium who transforms things and people. As a result, I think it is possible to draw a parallel between the procreative act of iron smelting, and the sexual potency of goats and smiths. All these have probably given the smiths significance in alleviating various social problems in the society. These were done both through their skills and in the medium of metal products and the animals they reared.

Beyond the ritual and medicinal significance of the animals reared by the smiths, one can see the question from the angle of economic benefits. Since smiths were not allowed to own large tracts of land for farming and rearing animals (see Hallpike 1968; Hultin 1977, 1984; Knutsen 1967), goats, which could be kept close to gardens, might have been their only choices. Smiths, through the help of ritual heads who themselves are related to them (Burka 2009), could have forged significance to smiths' animals so that they could extend their influence into various spheres of life of the society. Such a dimension must have helped the smiths to expand their economic basis beyond the meager income they obtained from metal production. The common myth of origin of the first smith and the first *Qaalluu* (Burka 2009) is meant to entrench the social basis of the smiths and the ritual heads. Since both sides complement each other, with the *tumtuu* forging ritual tools and the *Qaalluu* using the tools to alleviate problems by evoking their supernatural powers, the rest of the society is left with feelings of ambiguity about them.

The rest of the components of smelting (tuyères and furnaces) unlike the bag bellows are, despite various degrees of preservation, among metallurgical remains that form archaeological deposits. The cultural significance of the ethnoarchaeological record in iron smelting for archaeological interpretation is by presenting the various meanings attached to the body parts of the apparatuses. For instance, the furnace parts and extensions of southwest Wollega including the *faaliqaa* (*koshee*=vagina), the bulged-out- belly (symbolizing fertile woman), the two nipples on the 'chest', the 'mouth', ear-like projections on both sides of the 'head', the 'glottis' and 'epiglottis'-like shapes inside the mouth. These all parts not only form the anthropomorphic shape of the furnace but also evoke its feminine character. It is possible that a serious examination of archaeological finds of furnace remains in Wollega would yield results on 'unusual' body parts of the furnaces and thereby enable investigation of change and continuities in these cultural manifestations.

Similarly, the Wollega tuyères provide significant symbolic information. The clay *madabii* tuyère symbolizes an adult man's genitals (*qunxurroo*). The iron tuyère, on the other hand, is related to the inside of the genitals (*kololii*=channel). The third windpipe, horn *buudaa tuyère,* is meant to represent uncircumcised genital (*mundoo, buudaa*) and the bag bellows, to which the horn and iron tuyères are permanently attached symbolize the scrotum (*qalqala Cidhanii*).

Unlike the furnace, that clearly represents various sexual parts of a woman, the rest of the components of iron smelting stand for the male sexual parts of the body. On the one hand, this clearly suggests the procreative character of iron smelting. In the smelting setting, on the other hand, the smelters play the medium role of mediating the critical time of laboring of the pregnant woman (furnace) to deliver a child (the bloom=*dilalii*) and discharge the afterbirth (slag=*balxii*). Beyond the level of smelting, the meanings associated with the furnace (*gumbii*=granary), the bloom (*dilalii*=winnowed seed) and the slag (*balxxii, girdii*=chaff) invite us to see the connection between iron smelting and agricultural reproduction (*cf. chapter VI*). *Gumbii* (furnace) is particularly a woman phenomenon. It is not only that the furnace is called *gumbii* and that it is decorated as a woman, but also that in the ordinary use of the term *gumbii* represents a woman not a man. She is *gumbii* implies that the woman a fat one.

7.5. WOLLEGA EVIDENCE: THE WIDER CONTEXT

In order to place Wollega into a wider context of the study of iron metallurgy, one feature that archaeologists need to take seriously into consideration is its geographical position. Concerning geographical positions, I have earlier on (*cf. chapter I*) argued that the Ethiopia's geographical position *vis-à-vis* the rest of the surrounding world poses a serious challenge in explaining the sources of various facets of human achievements since antiquity. As a result, I have used Fattovich's (1996) *'Afro-Arabian circuit'* (emphasis is mine) model of cultural exchange on both sides of the Red Sea spanning time since the seventh millennium BC, to develop an explanatory model, which I preferred to call the *Third Model vis-à-vis* the advocates of external and internal origins' hypotheses.

One of the principles that guides the *Third Model* is the fact that no peoples either in the ancient past or in the present are 'closed and exclusive' to themselves. Consequently, the cultural manifestations of the historical period or at present should not have to blur the picture in favor of the view that the present carriers of certain knowledge are necessarily the creators themselves. Subscribing to such a view is tantamount to accepting the stagnancy of the ways of life of any society over a period or equally important is rejecting the mobile nature of humans.

Similarly, the Wollega case poses a challenge in view of population dynamics crossing and re-crossing the landscape since time immemorial leaving their impressions behind, which can be taken up and adapted to the new system by subsequent successors. In recent historical times, Wollega has served as a stage and bridge for the source as well as transfer of objects and ideas from and towards regions. Wollega has served as a bridge between regions north of the Blue Nile, across the western lowlands to the Sudan and south of the Gibe River to Oromo and Omotic-speaking inhabited regions of southwestern Ethiopia and eastwards to the center (Ahmad 1980, 1999; Burka 2006; Gidada 2001; Hassen 1990; Ta'a 1980; Triulzi 1973, 1975, 1981, 1996; Waqweyya 1988).

Despite this, the technological peculiarity of iron smelting apparatus in Wollega presents a difficulty of situating the evidence in a wider context. The southwestern Ethiopian Omo Valley smelting technology that employed between 6-10 'pot-bowl' bellows (see Haberland 1959: 425-26 and appendix; Haaland 2004a; Haaland *et al* 2000, 2004b; Todd 1985) operating on furnace superstructure built on a pit clearly shows the lack of any connection with the diversified Wollega furnaces. The number of tuyères Dime iron smelters used ranged from 30-60 (Haberland 1959; Haaland *et al* 2000, 2004b; Todd 1985) against 10 tuyères used by their Wollega counterparts.

In addition, the Omotic iron smelters made tuyères purely from clay as opposed to the three different materials of which the Wollega smelters and smiths made them. In terms of the depth of penetration by the clay tuyères deep inside the furnace, however, both regions show similarity, which could be accidental. As a result, a scenario that favors the dissemination of the knowledge from the southerly direction is less likely. An east west spread of the knowledge from central Ethiopia (Shoa) is also not plausible on similar technological grounds. Because of lack of research in the regions immediately north of the Blue Nile, no plausible suggestion is possible at this time.

The shoulder length vertical penetration of clay tuyères typical of southwestern Wollega can be comparable with many such types in western Africa (see Killick 1999; Sutton 1985; Tylecote 1980). Although contacts across the Sudanese belt is not unlikely, the fact that in West Africa (e.g. Cameroon) one single long tuyère was used with clay drum bellows (Killick 1999) emphatically distinguishes it from the Wollega material. As already noted, Wollega iron producers employed two long clay tuyères, four iron tuyères, four horn tuyères and the skin bag bellows. The pit furnace of northeast Wollega also lacks any comparison with its immediate neighbors.

On the other hand, whether the deep penetration of Wollega tuyères have any technological significances (see Schmidt and Avery 1987; Schmidt and Childs 1985; Schmidt 1997) needs a rigorous test under fully controlled laboratory environment (see Killick 1999). The use of over one meter long clay *madabii* (*Qunxurroo*) tuyères might have a cultural importance rather than having technological one.

7.3. CONCLUDING REMARKS

This study has sprung from an ambitious project where archaeology was unattempted before. The conviction was that everywhere archaeology has, at the beginning, started from scratch. Given this acute problem, an ethnoarchaeological approach, inspired by the *chaîne opératoire* framework, was adopted to attain the goal. Two study areas were selected. Two reasons explain for the selection of the sites namely that until about 40-50 years ago traditional iron smelting has continued in both areas, and that the practitioners in both areas claim common ancestral origin. In addition, the preliminary research result (Burka 2006, 2008) that showed the existence of distinct technologies of iron smelting called for examination of the reasons for variability. It was observed that the presence of technological variability among the Oromo society does not fit with the historical narrative about the people who were attributed to homogenization of the Ethiopian societies during the population movements of the 16th century (see Hassen 1990). On the other hand, previous researches on metallurgists in various Ethiopian regions (Cline 1937; Haberland 1959; Haaland *et al* 2004a, 2004b; Huntingford 1955; Lewis 1965; Todd 1985) could not clearly distinguish between iron smelters and the smiths and hence vestiges of respective production activities. In addition, in previous researches, we lack evidence regarding the nature of technological and socio-cultural aspects of iron metallurgy among the communities it was recorded.

The ethnoarchaeological study of iron metallurgy in Wollega has provided us with fresh insights towards the reconstruction of human past cultures in the region. It has shown that the migrationist model of technological diffusion is too simplistic to explain the complexities of developments in the region (see Burka 2006, 2008, 2009, 2010). Unlike in the literature where artisans were attributed to renegades who move around to disseminate knowledge, there is deep-rooted tradition among the Oromo of Wollega that both the mainstream and the artisans share common myth of origin. The ethnoarchaeometallurgical study has attested that this tradition can be distinctly identifiable on the archaeological landscape. It has clearly shown that iron smelters and smiths occupied different occupational identities-hence lived and worked in different localities-the products of which we can attribute to one clearly.

On the other hand, the study has shown the complexity of the problem and the need for future comprehensive research particularly regarding diverse technology among those who count close genealogy to the clan level. Two questions remain to be resolved: does the presence of technological variability indicate that it has been long experimented with and adapted by the people? On the other hand, does this variability indicate the origin of the knowledge from separate sources and periods? In addition, if the people did innovate iron smelting, then why did they innovate?

Nevertheless, since the study has attempted to address iron metallurgy in various socio-cultural settings of the Oromo people, it has yielded unrivaled information regarding the role of metallurgy and metals among the people. Metallurgy in the Oromo of Wollega was found to be integrated with the socio-economic and belief system of the people. In spite of its origin, the people seemed to have adapted it to their local knowledge system. On the one hand, the fertility of Women is invoked by the smelters in its production. Iron smelting procedure divides between female attributes from male ones: female and male ore, furnace as a female with anthropomorphic body parts while the tuyères (the horn, iron, and clay) and the bellows represent males' sexual organs. On the other hand, the fact that women (both granary maker and the potter) are credited with the origin of metallurgy and the metaphors in material culture (*gumbii, dilalii, balxii*) helps to go another level to situate the role of the knowledge in attaining agricultural productivity and sustenance of the society.

The fact that Wollega served as a stage for population movements and interactions fits with my third model that allows views regarding the forces that might govern the appearance and the distribution of certain knowledge from one region to another. As Wollega provides multitudes of major and minor streams overlooking the Nile Valley, future focus in this part of Ethiopia would securely yield archaeological data (including archaeometallurgy) for the purpose of regional and interregional analysis connected with exploitation of the various river valleys. A further future comprehensive research that should integrate various fields of study (including archaeological, anthropological, historical, linguistic, etc) would establish the culture-history of the region enabling comparisons among various communities inside and outside Wollega. In particular, a laboratory supported research of former iron production sites in Wollega and elsewhere in the country is an imperative to enlighten us with the status of early and late Iron Age cultures of Ethiopia and the Horn.

8. BIBLIOGRAPHY

Abdu, B, and Gordon, R. 2004. "Iron artifacts from the land of Kush" in *Journal of Archaeological Science* 31(2004), PP. 979-998.

Abir, M. 1966. "Salt, Trade and politics in Ethiopia in the 'Zamana Masafint' " in *Journal of Ethiopian Studies*, Vol. IV, No.2, 1-10.

Abir, M.1969. *Ethiopia: The Era of Princes: the challenge of Islam and the re-unification of the Christian Empire 1769-1855*. London: Longmans.

Agorsah, E.K. 1990. "Ethnoarchaeology: the search for a self-corrective approach to the study of past human behavior" in *African Archaeological Review*, 8:189-208.

Ahmad, Abdussamad H. 1980. "Trade and Politics in Gojjam 1882-1935". M.A. Thesis. Addis Ababa University.

Ahmad, Abdussamad H. 1999. "Trading in Slaves in Bela-Shangul and Gumuz, Ethiopia: Border Enclaves in history, 1897-1938" *The Journal of African History*, Vol. 40, and No.3:433-446.

Alpern, Stanley B. 2005. "Did They or Didn't They Invent It? Iron in Sub-Saharan Africa" in *Historyin Africa 32 (2005), 41-94.*

Anfray, F. 1981. "The Civilization of Aksum from the first to the seventh century." In Mokhatar, G. (ed.) *General History of Africa II: Ancient Civilizations of Africa*, University of California Press, Berkeley, PP. 362.380.

Arkell. J.A. 1966. "The Iron Age in the Sudan" *Current Anthropology* 7.

Arkell. J.A. 1968. (Rev.). *Meroe, a civilization of the Sudan*, by P.L. Shinnie, 1967, London: Thames and Hudson in *Journal of African History*, IX, 1 (1969).

Arnold III, J. Philip. 1991. *Domestic ceramic production and spatial organization: a Mexican case study in ethnoarchaeology*. New Studies in Archaeology. Cambridge University Press, Cambridge.

Atherton, J.H. 1983. "Ethnoarchaeology in Africa" in *African Archaeological Review*, 1: 75-104.

Bahrey, Abba.1954. "History of the Galla". In Beckingham, C.F. & B.G.W. Huntingford (Eds.) *Some Records of Ethiopia 1593-1646*. London. Hakluyt Society.

Bard, Kathryn A. and Fattovich, R. 2007. *Harbor of the Pharaohs to the Land of Punt: Archaeological Investigations at Mersa/Wadi Gawasis Egypt, 2001-2005*. Napoli 2007.

Barham, Lawrence and Mitchell, Peter. 2008. *The First Africans: African Archaeology from the earliest toolmakers to most recent foragers*. Cambridge University Press.

Barndon, R. 2004. "An Ethnoarchaeological Study of Iron-Smelting practices among the Pangwa and Fipa in Tanzania" in John Alexander and Lawrence Smith (Eds.)

Cambridge Monographs in African Archaeology, 61.

Bartels, Lambert. 1970. "Studies of the Galla in Wallaga: Their own views of the past" in *Journal of Ethiopian Studies*. Vol. VIII, No.1.

Bartels, Lamber. 1983. *Oromo Religion: Myths and Rites of the Western Oromo of Ethiopia: an attempt to understand*. Dietrich Reimer Verlog. Berlin.

Baxter, P.T.W. 1978. "Ethiopia's unacknowledged problem: The Oromo" in *African Affairs*, Vol. 77, No. 308.

Baxter, P.T.W. and Blackhurst, H. 1978. "Vercingetorix in Ethiopia: Some problems arising from Levine's inclusion of the Oromo in his delineation of Ethiopia as a culture area" *Abbay* 9: 159-166.

Bent, T. 1896. *The Sacred City of the Ethiopians: Being a Record of Travel and Research in Abyssinia in 1893*. London: Longmans, Green and Co.

Beyin, Amanuel and Shea, John J. 2007. "Reconnaissance of prehistoric sites on the Red Sea coast of Eritrea, NE Africa" in *Journal of Field Archaeology* Vol. 32 No.1, 2007.

Blackhurst, Hector. 1996. "Adopting an Ambiguous position: Oromo Relationships with Strangers" In P.T.W. Baxter, Jan Hultin, and Alessandro Triulzi (eds.) *Being and Becoming Oromo: Historical and Anthropological Enquiries*. Nordiska Afrikaninstitutet, Uppsala.

Brandt, S.A. et al. 1997. "Gurage hide working, stone tool use and social identity: an ethnoarchaeological perspective", in G. Hudson (ed.), *Essays on Gurage Language and Culture dedicated to Wolf Leslau*. Wiesbaden: Harrassowitz Verlag.

Brandt, S.A., and Fottovich, R. 1990. "Late Quaternary Archaeological Research in the Horn of Africa" in Peter Robertshaw (ed.) *A History of African Archaeology*. James Currey: London: 95-108.

Burka, Temesgen. 2004. "The Role of Museums and Galleries in Ethiopia". MA Thesis, College of Social Sciences, Addis Ababa University.

Burka, Temesgen. 2006. "Iron Smelting in Wollega, Ethiopia: An Ethnoarchaeological Study". MA Thesis, University of Bergen, Norway.

Burka, Temesgen. 2008. "The Socio-cultural aspects of furnace shape, decorations: Iron smelting activities among the Oromo of West Wollega, Ethiopia" in *Nyame Akuma*, N. 69, June 2008. pp.1-9.

Burka, Temesgen. 2009. "Why the Tumtuu are Not Marginalized: New Perspectives on the Positions of the Smiths in Western Ethiopia, Wollega" in *Nyame Akuma*, N.72. December 2009. pp.12- 20.

Burka, Temesgen. 2010. "Cultural Aspects of Metallurgy in Ethiopia: the case of Iron productionin Wollega" In Claudio Giardino (ed.) *Archaemetallurgia dalla conoscerza alla fruizione*, Congedo editor, Galatina (Lecce).

Casson, L. 1989. *Periplus Maris Erythraei*, Princeton University Press, Priceton, NJ.

Cerulli, E. 1933. *Etiopia Occidentale: Dallo Scio Alla Frontiera Del Sudan*. Volume Secondo, Sindacato Italiano Arti Grafiche, Rome, 1933-XI

Cerulli, E. 1922.*The Folk-Literature of the Galla*. Harvard African Studies, Varia Africana III, 9-228. Cambridge MA: Harvard University press.

Childs, S. Terry and Herbert Eugenia W. 2005. "Metallurgy and its Consequences" in Ann Brower Stahl (ed.) *African Archaeology: A Critical Introduction*. Blackwell Publishing. PP.276-300.

Childs, S. Terry and Killick, D. 1993. "Indigenous African Metallurgy: Nature and Culture" *Annual Reviews of Anthropology* 22: 317-337.

Chirikure, S. 2005. "Iron Production in Iron Age Zimbabwe: Stagnation or Innovation?" Ph D Thesis. University of London.

Chirikure, S.2007. "Metals in Society: Iron Production and its position in Iron Age communities of Southern Africa" in *Journal of Social Archaeology, Vol. 7 (1): 72-100.*

Chittick, H.N. 1978. " Notes on the archaeology of Northern Ethiopia" in *Abbay*, 9: 15-19. Chittick, H.N. 1974. "Excavations at Aksum 1973-74: A preliminary report" in *Azania* 9: 159-205. Clark, J.D. *Tools and Ourselves: an African legacy* (seventeenth Raymond Dart Lecture). Johannesburg: Institute for the Study of Man.

Cleere, H. 1972. "The classification of early iron smelting furnaces" in *Antiquaries Journal* 52:8-23.

Cleere, H. 1997. (Rev.) *Early metal mining and production* by Paul T. Craddock, 1995. In *Journal of Field Archaeology* Vol.24, No.1, 2007.

Cline, W. 1937. *Mining and Metallurgy in Negro Africa*. Menasha: George Banta. Cole, S. 1954. *The Prehistory of East Africa*, 2nd ed. Penguin, Harmondsworth.

Collett, D.P. 1993. "Metaphors and representations associated with pre-colonial iron smelting in eastern and southern Africa" in Shaw, T., Sinclair, P., Andah, B. Okpoko, A. (Eds.) *The archaeology of Africa: food, metals and towns*. London: Routledge.

Contenson, de H. 1981. "Pre-Aksumite culture" in Mokhtar, G. (ed.) *General History of Africa II: ancient civilizations of Africa*. University of California Press. Berkeley. PP.341- 361.

Costin, C.L. "Thinking about Production: Phenomenological Classification and Lexical Semantics" in *Archaeological papers of the American Anthropological Association* Vol.17 (1):143-162.

Cresswell, R. 1990 " 'òA New Technology' revisited", *Technology and the Humanities, Archaeological Review from Cambridge*, 9(1), pp. 39-54.

Crew, Peter. 1995. "Bloomery Iron Smelting and other residues" in *The Historical Metallurgy Society Archaeology Datasheet No.5.*

Crew, Peter. 1996. "Bloom refining and smithing slags and other residues" in *The Historical Metallurgy Society Archaeology Datasheet No. 6.*

Crummey, D. 1991 (Rev.) "Civilization at Aksum: Excavations at Aksum: An account of research at ancient Ethiopian capital directed in 1972-1974 by the late Dr. Neville Chittick" by S.C. Munro-hay, D.W. Phillipson (ed.) in *The Journal of African History*, Vol. 32 (1):144-146. Cambridge University Press.

Crummey, D. 1990. "Society, State and Nationality in the Recent Historiography of Ethiopia" in *The Journal of African History* 31:103-119.

Curtin, Philip D. 1968. "Field Techniques for Collecting and Processing Oral Data" in *Journal of African History IX, 3* (1968): 367-385.

David, Nicholas. 2001. "Lost in the third Hermeneutic? Theory and Methodology of African Metallurgy" in *Mediterranean Archaeology* 14. 49-72.

David, N. R. Heimann, D. Killick, and M. Wayman. 1989. "Between bloomery and blast furnace: Mafa iron-smelting technology in northern Cameroon". *The African Archaeological Review*,7: 183-208.

David, N. and Kramer, C. 2001. *Ethnoarchaeology in Action*. Cambridge World Archaeology. Cambridge University Press. pp. 410-18.

Dea, D. 2003. "The challenge of integrative Power: Hierarchy and Political change in Dawro, southern Ethiopia". Ph D. thesis, University of Bergen.pp.26-43.

De Barros, Phillip. 2000. « Iron Metallurgy: Sociocultural context ». In Bisson, M, Childs, T, De Barros, P, and Holl, A. (eds.) *Ancient African Metallurgy: The Socio-cultural Context*. New York: Altamira Press. 147-199.

De Barros, Phillip. 1997. "Iron working in its cultural context". In Joseph, O.Vogel. (ed.) *Encyclopedia of Pre-colonial Africa: Archaeology, History, Languages, Cultures and Environments*, (PP. 135-149). Walnut Creek: AltaMira.

Diop, C. A. 1974. *The African Origin of Civilization: Myth or Reality*, Lawrence Hill Books, Chicago.

Diop, C. A. 1981. *Civilization or Barbarism: An Authentic Anthropology*, Lawrence Hill Books, Chicago

Durrani, N. 2005. *The Tihamah Coastal plain of south-west Arabia in its regional context c.6000 BC- AD 600*. BAR International Series 1456. Oxford: Archaeopress Edwards, David N. 1996. The *Archaeology of the Meroitic State: New perspectives on its social and political organization*. BAR International Series 640.

Ehret, Christopher. 1997. "African Languages: A Historical Survey". In Joseph, O. Vogel. (ed.) *Encyclopedia of Pre-colonial Africa*: *Archaeology, History, Languages, Cultures and Environments*, (PP. 159-166). Walnut Creek: AltaMira.

Ehret, Christopher. 1982. "Population movement and cultural contact in the southern Sudan, c. 3000 BC to AD 1000: A preliminary linguistic overview". In Mack, J., and Robertshaw, P. (eds.) *Culture History in the southern Sudan*. BIEA, Nairobi: PP. 19-48.

Ehret, Christopher. 1973. "Patterns of Bantu and Central Sudanic settlement in Central and Southern Africa (ca. 1000BC-500AD)". In *Transafrican Journal of History*, 3: 1-71.

Fattovich, R. 1978. "Traces of a possible African component in the Pre-Aksumite culture of northern Ethiopia" in *Abbay* 9:25-30.

Fattovich, R. 1990. "Remarks on the Pre-Aksumite period in northern Ethiopia" in *Journal of Ethiopian studies Vol. XXIII*: 1-33. Addis Ababa.

Fattovich, R. 1997a. "The Contacts between Southern Arabia and the Horn of Africa in late prehistoric and early historic times: a view from Africa" in A. Avanzini (ed.), *Profumi d'Arabia* (pp. 273-286). Rome: 'L'Erma' di Bretschneider.

Fattovich, R.1997b. "The Near East and Eastern Africa: their interaction" in Joseph, O. Vogel. (ed.) *Encyclopedia of Pre-colonial Africa*: *Archaeology, History, Languages, Cultures and Environments*, (PP. 479-484). Walnut Creek: AltaMira.

Fattovich, R.1996. "The Afro-Arabian circuit: Contacts between the Horn of Africa and Southern Arabia in the 3rd- 2nd millennium BC". In L. Krzyzania, K. Kroeper and M. Kobusiewick (eds.) *Interregional Contacts in the Later Prehistory of Northeastern Africa*, 395-402. Poznàn: Poznàn Archaeological Museum. Fattovich, R. 2009. "Reconsidering Yeha, c. 800-400 BC" in *African Archaeological Review*, 26 (4): 275- 290.

Fattovich, R. *et al.* 2000. "Aksum and the Habashat: state and ethnicity in ancient northernEthiopia and Eritrea"in *Working Papers in African Studies* 228. Boston MA: Boston University African Studies Center.

Fattovich, R. and Bard, Kathryn A. 1997. "The I.U.O. /B.U. Excavations at Beta Giyorgis (Aksum) in Tigray (Northern Ethiopia)" in *Journal of Ethiopian Studies.* Vol. XXX. No.1

FDRE. 1996. *The 1994 Population and Housing Census of Ethiopia: Results for Oromiya Region, Volume I: Part I Statistical Report on Population Size and Characteristics.* FDRE: Office of Population and Housing Census Commission, CSA. Addis Ababa (1996) pp.184-187.

Filipowiak, W.1985. "Iron working in the old kingdom of Mali" in Haaland, R. and Shinnie, P. (eds.) *African Iron Working: Ancient and Traditional.* Norwegian University Press. New York.

Finneran, N.2007. *The Archaeology of Ethiopia.* Routlage: London and New York.

Fitzhugh, Ben. 2001. "Risk and Innovation in Human Technological Evolution" in *Journal of Anthropological Archaeology* 20 (2001).

Fleming, H. 1976. 'Sociology, Ethnology, and History in Ethiopia' in *The International of African Historical Studies*, IX, 2:248-278.

Freeman, Dena. 2001. "Conclusion I" in Freeman, D and Pankhurst, A. (eds.) *Living on the Edge: Marginalized Minorities of Craft workers and Hunters in Southern Ethiopia.* Addis Ababa University Press.

Gallay, Alain. 2001. « Diffusion ou Invention: Un faux debat pour l'archeologie » in *Mediterranean Archaeology* 14. 13-24.

Gidada, Negaso. 2001. *History of the Sayyo-Oromo of southwestern Wollega, Ethiopia from about 1730-1886.* Frankfurt am Main.

Giumlia-Mair, Alessandra. 2002. "Investigation of a copper based hoard from the Megalithic site of al-Midamman, Yemen: An interdisciplinary Approach" in *Journal of Archaeological Science* 29(2002).

Gould, R.A. 1978. "Beyond analogy in Ethnoarchaeology" in R.A. Gould (ed.) *Explorations in Ethnoarchaeology.* Albuquerque: University of New Mexico Press.

Greenfield, H.J. and Miller, D. 2004. "Spatial patterning of Early Iron Age metal production at Ndondondwane, South Africa: the question of cultural continuity between the Early and Late Iron Ages" in *Journal of Archaeological Science* 31(2004).

Haaland, R. 2007. "Say it in Iron: Symbols of Transformation and Reproduction in the European Iron Age" in *Current Swedish Archaeology,* Vol. 15, 2007:1-20.

Haaland, R. 2004a. "Technology, Transformation, and Symbolism: Ethnographic perspectives on European iron working" in *Norwegian Archaeological Review* 37(1).

Haaland, R. 2004b. "Iron smelting – A vanishing tradition: Ethnographic Study of this craft in Southwest Ethiopia" in *Journal of African Archeology* 2 (1).

Haaland, R. 1988. "The role of ethnoarchaeology and experimental archaeology in the

interpretation of prehistoric societies" in *Nordic Archaeological Abstracts* 67:130-139. Haaland, R. 1985. "Iron production, its socio-cultural context and ecological

Implications" in Haaland, R., and Shinnie, P. (eds.) *African Iron Working:Ancient and Traditional.* Norwegian University Press. New York.

Haaland, R. 1980. "Man's role in the changing habitat of Mema during the Old Kingdom of Ghana" in *Norergian Arhcaelogical Review 13.* 31-46.

Haaland, R. and Haaland, G. 2008. "Craft Specialization, Caste identities and Political centralization: On the use of Anthropological perspectives in reconstructing archaic forms of economic organization" in *OAS Nr.10. Facets of Archeology.* Essays in Honor of Lotte Hedeager on her 60th birthday.

Haaland, G. and Haaland, R. 2007. "God of war, worldly ruler, and craft specialists in the Meroitic Kingdom of Sudan: Inferring social identity from material remains" in *Journal of Social Archaeology* Vol. 7(3):372-392.

Haaland, G., Haaland, R., and Dea D. 2000. "Ethnoarchaeological research on iron smelting in southwest Ethiopia" in *Nyame Akuma* 54:6-12

Haaland, G., Haaland, R., and Dea D. 2004a. "Smelting Iron: Caste and its symbolism in South-Western Ethiopia" in T. Insoll (ed.), *Belief in the past: The Proceedings of the Manchester Conference on Archaeology and Religion.*

Haaland, G., Haaland, R., and Dea D. 2004b. "Furnace and Pot: why the iron smelter is a big pot maker: A case study from south-western Ethiopia" in *Azania* XXXIX 2004.

Haaland, G., Haaland, R., and Rijal, S. (2002). "A cross-cultural study of technological, symbolic, and social aspects of iron making" in *Anthropos*.97.

Haberland, E.1959. *Altvolker Sud-Ethiopiens. Band I.* Stuttgart: W. Kholhammer Verlag.
Haberland, E. 1963. *Galla Sud-Athiopiens.* Stuttgart: Kohlhammer Verlag.
Haberland, E. 1978. "Special castes in Ethiopia", in R. Hess (ed.) *Proceedings of the Fifth International Conference of Ethiopian Studies, Session B* PP. 129-132. Chicago: University of Chicago Press.
Haberland, E. 1984. "Caste and hierarchy among the Dizi (southwest Ethiopia)", in S. Rubenson (ed.) *Proceedings of the Seventh International Conference of Ethiopian Studies*, PP. 447-450. Addis Ababa: Institute of Ethiopian Studies
Hall, M. 1987. *The Changing Past: Farmers, Kings and Traders in Southern Africa, 200-1860.* David Philip. Cape Town & Johannesburg.
Hallpike, C.R. 1968. "The Status of Artisans among the Konso of Southwest Ethiopia" in *Africa: Journal of the International African Institute*, Vol. 38, No.3.
Hamrla, M. 1966. "The Iron and Manganese Ore Deposits in Ethiopia" in *Geologija,* 9:439-487.
Hamrla, M. 1964. "Report on Journey to Wollega", Geological Department, Ministry of Mines.
Hammersley, M. and Atkinson, P. 1995. *Ethnography: principles in practice.* Second Edition, London and New York, PP. 1-122, 157-174.
Hassen, Mohammed. 1990. *The Oromo of Ethiopia: A History, 1570-1860.* Cambridge University Press
Herbert, E.W. 2001. "African Metallurgy: The Historian's Dilemma" in *Mediterranean Archaeology 14.*
Herbert, E.W. 1993. *Iron, Gender and Power: Rituals of Transformation in African Societies.* Bloomington: Indiana University Press.
Hodder, Ian. 2003. "Archaeological Reflexivity and the 'Local' voice" in *Anthropological Quarterly*, Vol. 76, No. 1.
Hoddeer, Ian. 1986. "From Ethnoarchaeology to Material Culture Studies" in *Dialoghi Di Archeologia*, Terza Serie, No.1 PP.95-97.
Holl, A. 2008. "Early West African Metallurgies: Glosses, Facts and Fictions" in *SAA 2008, Canada, Session: 'Modeling Early Metallurgy: Old and New Perspectives'*, Vancouver, March 26-28, 2008.
Holl, A.1997. "Pan-Africanism, Diffusionism, and the Afrocentric Idea" in Joseph O. Vogel (Eds.). *Encyclopedia of precolonial Africa: archaeology, history, languages, cultures, and environments.* AltaMira Press. Pp. 58-65.
Hultin, J. 1984. "Sons of slaves or sons of boys: On the premise of rank among the Macha Oromo" in Beyene, T. (ed). *Proceedings of the Eighth International Conference of Ethiopian Studies.* Huntingdon.
Huntingford, G.W. 1931. "Free hunters: serf-tribes and submerged classes in East Africa", in *Man*, Vol. 31, No.261, PP.261-266.
Huntingford, G.W. 1955. "The Galla of Ethiopia. The Kingdom of Kafa and Janjero" in Daryil Forde (eds.) *Ethnographic Survey of Africa.* London.
Irvine, A.K. 1978. "Linguistic Evidence on Ancient Ethiopia: The relationship of early Ethiopian Semitic to old south Arabian" in *Abbay*, 9:43-48.
Jalata, A. 1996. "The Struggle for knowledge: The case of Emergent Oromo Studies" in *African Studies Review*, Vol.39, No.2, pp. 95-123.
Jensen, A. 1959. *Altvolker Sud-Athiopiens.* Stuttgart: W. Kohlhammer Verlag.
Kanno, Ayalew. 2007. *On the Oromo: Great African Nation Often Designated Under the Name 'Galla'.* Antoine D'Abbadie, Translation from French.
Kassam, A. 1999. "Ritual and classification: A study of the Booran Oromo Terminal Sacred grade Rites of Passage" in *Bulletin of the School of Oriental and African Studies.* University of London, Vol. 62, No.3. PP.484-503.
Kassam, A. and Bushana, Ali Balla. 2004. Marginalization of the Waata Oromo Hunter-Gatherers of Kenya: Insider and outsider perspectives. *Africa: Journal of the International African Institute*, 74 (2):194-216.
Kassam, A. and Megerssa, G. 1989. "Iron and Beads: Male and female symbols of Creation: A study of ornament among Booran Oromo" in Hodder, I. (Ed). *The Meaning of Things. Material culture and symbolic expression.* London, Unwin Hyman Ltd.
Kense, F.J. 1985. "The initial diffusion of iron to Africa" in Haaland, R. and Shinnie, P. (Eds.) *African Iron Working: Ancient and Traditional.* Norwegian University Press, New York. Kense, F.J. and Okoro,

J. Ako. 1993. "Changing Perspectives on traditional iron production in West Africa" in T., Sinclair, P. Andah, B., and Okpoko, A. (eds.)*The Archaeology of Africa: Food, Metals and Towns*. Routledge. London.

Khan, F. 1994. "The potential of ethnoarchaeology with special reference to recent archaeological work in Banu district, Pakistan" in *Living Traditions: Studies in the archaeology of South Asia*, PP. 83-89, New Delhi: Oxford and IBH Publishing Co. Pvt. Ltd.

Killick, David. 1999. (Review of). "Iron Technology in East Africa: Symbolism, science and Archaeology by Peter R. Schmidt (1997)" in *Journal of Field Archaeology*, Vol.26 No.3, 1999.

Killick, David. 2004. "What do we know about African iron working?" in *Journal of African Archaeology*. Vol. 2(1).

Killick, David. 2009. "Cairo to Cape: The Spread of Metallurgy Through Eastern and Southern Africa" in *Journal of World Prehistory* (2009) 22:399-412.

Knutson, K.E. 1967. *Authority and Change: A study of the Kallu institution among the Macha Galla of Ethiopia*. Gotenborg, Ethografiska Museot.

Kobylinski, Zbigniew. 1989. "Ethno-archeological Cognition and Cognitive ethno-archaeology" in Hodder, I. (ed). *The Meaning of Things. Material culture and Symbolic expression*. London, Unwin Hyman Ltd.

Kramer, C. 1985. " Ceramic ethnoarchaology" in *Annual Review of Anthropology* 14: 77-102. Kurimoto, Eisei. 1995. "Trade Relations between western Ethiopia and the Nile Valley during the nineteenth century" in *Journal of Ethiopian Studies*. Vol. XXVIII, NO.1. 55-68. Legesse, A. 1973. *Gada: Three Approaches to the Study of African Society*. New York, Free Press. Levine, Donald N. 1974. *Greater Ethiopia, the Evolution of a Multiethnic Society*. Chicgo University Press.

Lewis, H.S. 1962. "Historical problems in Ethiopia and the Horn of Africa" *in Annals of the New York Academy of Science* 6:504-511.

Lewis, H.S. 1965. *A Galla Monarchy: Jimma Aba Jifar, Ethiopia (1830-1932)*. Madison: University of Wisconsin.

Lewis, H.S. 1970. "Wealthy, influence and prestige among the Shoa Galla", in A. Tuden and L.

Plotnicov (eds.) *Social Stratification in Africa*, PP. 163-189. New York and London: Free Press and Collier-Macmillan.

Lewis, H.S. 1984. "Oromo", in Richard V. Weeks (ed.) *Muslim peoples: A world Ethnographic survey*. Second Edition. Greenwood Press.

London, Gloria. 2000. "Ethnoarchaeology and Interpretation of the past" in *Near Eastern Archaeology*, Vol.63, No.1, PP.2-8.

Longacre, W.A. 1991(ed.) *Ceramic Ethnoarchaeology* . Tucson: University of Arizona Press

Magee, P. 1996. "The Chronology of Southeast Arabian Iron Age" in *Arabian Archaeology and Epigraphy* 7.

Manzo, A. "*Capra nubiana* in *Berbere* Sauce? Pre-Aksumite Art and Identity Building" in *African Archaeological Review*, 26(4): 291-303.

Mapunda, B.B. 1997. Patching up evidence for iron working in the Horn. *African Archaeological Review* 14(2).

Mapunda, B.B. 2000. (Review of) *The culture and technology of African Iron production*, (1996) Peter Schmidt (ed.) and *Iron Technology in East Africa: Symbolism, Science and Archaeology* by Peter Schmidt (1197) in *Journal of African History* 41(2000), Pp. 487-89.

Martinon-Torres, M. 2002. "Chaîne opèratoire: The Concept and its Applications within the Study of Technology" in *Gallaecia*, 21:29-43.

Mbaya, Maweja. 2002. "Linguistic taboo in African marriage context: A study of the Oromo *Laguu"* in *Nordic Journal of African Studies*, 11(2).

Mc Donnell, Gerry. 1995a. "Iron working processes" in *The Historical Metallurgy Society. Archaeology Datasheet No.3.*

Mc Donnell, Gerry.1995b. "Geophysical techniques applied to early metal working sites"in *The Historical Metallurgy Society. Archaeology Datasheet No.4.*

Mc Donnell, G. and Stanley, D. 2002. "Excavation and sampling strategies for metal working Sites"in *The Historical Metallurgy Society Ltd. Archaeology Datasheet No.9*

Mcintosh, Roderick J. 2005. *Ancient Middle Niger: Urbanism and the Self-Organizing Landscape*. Cambridge University Press.

Michels, J. 1990. "Review Article: Excavations at Aksum". In *African Archaeological Review* 8: 177-188.

McCosh, F.W.J. 1979. "Traditional iron-working in central Africa with some references to the ritualistic and scientific aspects of the industry". In *Zambezia*, VII (ii): 155-170.

Megerssa, G. 1994. "Knowledge, Identity and the Colonizing Structure: the case of the Oromo in East and Northeast Africa". University of London. Ph D Thesis.

Miller, Duncan. 2002. "Smelter and Smith: Iron Age metal fabrication technology in Southern Africa" in *Journal of Archaeological Science* 29, 1083-1131.

Miller, Duncan.2001. "Indigenous Iron production in southern Africa: Archaeological observations and interpretation" in *Mediterranean Archaeology 14*, 229-234.

Miller, D. 1997. "Iron Working Technology". In Joseph, O.Vogel. (ed.) *Encyclopedia of Pre-colonial Africa: Archaeology, History, Languages, Cultures and Environments*, (PP. 132-135). Walnut Creek: AltaMira.

Miller, D. and Van der Merwe, N. 1994. "Early Metal working in Sub-Saharan Africa: A review of recent research" in *Journal of African History* 35: 1-36.

Mitchell, Peter. 2005. *African Connections: Archaeological Perspectives on Africa and the Wider World*. ALTAMIRA PRESS.

Muhly, J.1980. "The Bronze Age setting". In Wertime, T., and Muhly, J. (eds.) *The Coming of the Age of Iron*. Yale University Press, New Haven, CT. and London, PP. 25-67.

Munro-Hay, S. 1993. "State development and urbanism in northern Ethiopia" inT, Sinclair, P. Andah, B., and Okpoko, A. (eds.) *The Archaeology of Africa: Food, Metals and Towns*. Routledge. London.

Munro-Hay, S. 1991. *Aksum: an African civilization of late antiquity*, Edinburgh University Press, Edinburgh.

Murdock, George. 1959. *Africa: its peoples and their culture history*. New York: McGraw-Hill

Murray, G.W. and Warmington, E.H. 1967. "Trogodytica: The Red Sea littoral in Ptolemaic Times" in *The Geographical Journal*, Vol. 133, No.1.

Musa-Mohamed, Ibrahim. 2004. "Traditional Iron technology and settlement patterns in central Darfur" in *Azania* XXXIX 2004.

Nadel, S. 1954. "Caste and government in primitive society" in *Journal of the Anthropological Society of Bombay*, Vol. 8, No.2, PP. 9-22.

Okafor, E.E. 1993. "New evidence on early iron-smelting from southeastern Nigeria" inShaw, T., Sinclair, P., Andah, B. Okpoko, A. (eds.) *The archaeology of Africa food, metals and towns*. London: Routledge.

Oliver, R. and Fagan, B. 1975. *Africa in the Iron Age; c. 500BC to AD 1400*. Cambridge University Press. Cambridge.

Pankhurst, A. 1999. "Caste in Africa: the evidence from southwestern Ethiopia Reconsidered" in *Africa* Vol. 69, No. 4, pp.485-509.

Pankhurst, A. 2001. "Introduction" in Freeman, D. and Pankhurst, A. (eds.) *Living On the Edge: Marginalized Minorities of Craft workers and Hunters in Southern Ethiopia*. Addis Ababa University Press.pp.1-23.

Pankhurst, R. 1961. "Status, division of labor and employment in nineteenth century and early twentieth century Ethiopia". In *Ethnological Society Bulletin*, Vol. II, N. 1, University College of Addis Ababa.

Pfaffenberger, B. 1992. "Social Anthropology of Technology". In *Annual Review of Anthropology, vol. 21*.

Phillipson, D.W.1977. "The Excavation of Gobedra Rock Shelter, Axum: an early occurrence of cultivated finger millet in Northern Ethiopia" in *Azania*, XII. Journal ofthe British
Institute in Eastern Africa.

Phillipson, D.W. 1990. "Aksum in Africa" in *Journal of Ethiopian Studies, Vol. XXIII*: 55-65. Addis Ababa.

Phillipson, D.W. 1993. "The antiquity of cultivation and herding in Ethiopia" in T., Sinclair, P. Andah, B., and Okpoko, A. (eds.) *The Archaeology of Africa: Food, Metals and Towns*. Routledge. London.

Phillipson, D.W. 1995. "The B.I.E.A Aksum Excavations, 1994." In *Nyame Akuma*, 44:28-34. Phillipson, D.W. 2005. *African Archaeology*. Third Edition. Cambridge University Press. Phillipson, D.W. 2009. "The First Millennium BC in the Highlands of Northern Ethiopia and
South-Central Eritrea: A Reassessment of Cultural and Political Development". *African Archaeological Review*, 26 (4): 257-274.

Pleiner, R. 2000. *Iron in Archeology: The European bloomery smelters*. Praha. Pleiner, R. 1980. "Early Iron Metallurgy in Europe" in Wertime and Muhly (eds.) *The coming of the Age of Iron*. New Haven. Yale University Press. 375-415.

Posnansky, M. 2009. *Africa and Archaeology: Empowering an Expatriate Life*. The Radcliffe Press:London. New York

Posnansky, M. 1966. "The origins of Agriculture and Iron-working in southern Africa". In

Posnansky, M. (ed.) *Prelude to East African History*, Oxford University Press, London, PP.82-94.

Potter, James M. 2000. (Review of) *Metaphor and Material Culture* by Christopher Tilley (1999). Blackwell Publishers Ltd. In *Journal of Field Archaeology* Vol. 27 No.1.

Quirin, James. 1997. "Society and the state: reflections from the northwest, 1300-1900" in

K. Fukui, E. Kurimoto and M. Shigeta (eds.) *Ethiopia in Broader Perspective: Papers of the Thirteenth International Conference of Ethiopian Studies*. Kyto, Shokado Booksellers. Reid, Andrew and MacLean, Rachel. 1995. "Symbolism and the social contexts of Iron production in Karagwe" in *World Archaeology*, Vol. 27, No.1

Rikitu, Mengesha. 2001. *The Oromo of the Horn: Cultural History, Past and Present*. Brook Green Print-London.

Rothenberg, Beno. 1995. "Archaeometallurgical survey method" in *Proceedings of the First International Conference on Ancient Egyptian Mining and Metallurgy and Conservation of Metallic Artifacts*, Egypt Cairo, 10-12 April 1995.

Rowlands, M. and Warnier, J. P.1993. "The magical production of iron in the Cameroon Grass fields" in Shaw, T., Sinclair, P., Andah, B. Okpoko, A. (eds.) *The archaeology of Africa: food, metals and towns*. London: Routledge.

Sayce, A. H. 1912. "Second Interim Report on the excavations at Meroe in Ethiopia". *Liverpool Annals of Archaeology and Anthropology* 4: 53-65.

Schiffer, M.B. 1978 "Methodological Issues in Ethnoarchaeology" in R.A.Gould (ed.) *Explorations in Ethnoarchaology*. Albuquerque: University of New Mexico Press.

Schlanger, N. 1994. 'Mindful technology: unleashing the chaîne opératoire for archaeology of mind' in C. Renfrew and E.B.W. Zubrow (eds.) *The Ancient Mind: Elements of cognitive archaeology*. (New Directions in Archaeology). Cambridge: Cambridge University Press, PP.143-151.

Schmandt-Besserat, Denise. 1980. "Ocher in Prehistory: 300,000 Years of the Use of Iron Ores as Pigments" in Wertime and Muly (eds.) The Coming of the Age of Iron. New Haven. Yale University press.127-150.

Schmidt, P.R. 2009. "Variability in Eritrea and the Archaeology of the Northern Horn during the First Millennium BC: Subsistence, Ritual, and Gold Production" in *African Archaeological Review*, 26 (4):305-325.

Schmidt, P. R. 2001. "Resisting homogenization and recovering variation and innovation inAfrican iron smelting" in *Mediterranean Archaeology 14*. 219-227.

Schmidt, P. R. 1997. *Iron technology in East Africa: Symbolism, Science and Archaeology*. Oxford:James Currey.

Schmidt, P. R. and Avery, D. 1978. "Complex iron smelting and prehistoric culture in Tanzania" in *Science* 201, 1085-1089.

Schmidt, P. R. and Avery, D.1983. "More evidence for an advanced Prehistoric Iron Technology in Africa" in *Journal of Field Archaeology* 10:421-434.

Schmidt, P. R. and Childs, T. S. 1985. "Innovation and Industry during the Early Iron Age in East Africa: the KM2 and KM3 sites of Northwest Tanzania"in *The African Archaeological Review* 2.

Schmidt, P. R. and Mapunda, B. B. 1997. "Ideology and the Archaeological record in Africa: Interpreting Symbolism in iron smelting technology" in *Journal of Anthropological Archaeology, 16.*

Selassie, S.M. 1972. *Ancient and Medieval Ethiopian History to 1270*. Addis Ababa.

Shack, W. 1964. "Notes on occupational castes among the Gurage of south-west Ethiopia" in *Man* 54:50-52.

Shepherd, N. 2002. "The Politics of Archaeology in Africa". In *Annual Review of Anthropology*, 31: 189-209.

Shinnie, P. 1967. *Meroe, a civilization of the Sudan*. Thames and Hudson. London. Shinnie, P. 1985. "Iron working at Meroe". In Haaland, R. and Shinnie, P. (eds.) *African Iron Working: Ancient and Traditional*. Norwegian University Press, New York.

Simoons, F. 1965. "Some Questions on the economic prehistory of Ethiopia" in *Journal of African History* 6:1-13.

Sinclair, Paul, Shaw, Thurstan and Andah, Bassey. 1993. "Introduction". In Thurstan Shaw *et al* (eds.) *The Archaeology of Africa: Food, metals and towns.* Routledge. London and New York. PP.1- 32.

Stahl, Ann Brower. 2005. "Introduction: Changing perspectives on Africa's pasts" in Ann Brower Stahl (ed.) *African Archaeology: A critical Introduction.* Blackwell Publishing Ltd.

Stanislawski, M.B. 1978. "If pots were mortal" in *Explorations in Ethnoarchaeology*, R.A. Gould (ed.) PP. 201-228. Albuquerque: University of New Mexico Press

Stanley, David. 1995. "Hammerscale" *The Historical Metallurgy Society. Archaeology Datasheet No.10*

Staski, E, and L.D. Sutro (eds.) *The ethnoarchaeology of refuse disposal.* Anthropological Research Papers, 42. Tempe: Arizona State University

Sutton, J.E.G. 1985. "Temporal and Spatial Variability in African iron furnaces" in Haaland, R. and Shinnie, P. (eds.) *African Iron Working: Ancient and Traditional.* Norwegian University Press, New York.

Ta'a, Tesema. 1980. "The Oromo of Wollega: a Historical Survey to 19102". MA Thesis, Addis Ababa University.

Ta'a, Tesema, 1996, Traditional and modern cooperatives among the Oromo. In, Baxter, P.T.W., Hultin, J., and Triulzi, A. (eds.) *Being and Becoming Oromo: Historical and Anthropological Enquiries.* Nordiska Afrikainstitutet, Uppsala.

Tamrat, Taddesse. 1970. "The abbots of Dabra Hayq, 1248-1535" in *Journal of Ethiopian Studies.* Vol.8/1: 87-117.

Tamrat, Taddesse. 1972. *Church and State in Ethiopia: 1270-1527.* The Clarendon Press. Oxford Thornton, C.P. and Giardino, C. 2008. "Alla ricerca pardigmo archaeometallurgico" in *Rivista di Scienza Preshistoriche*-LVIII, PP.385-404.

Todd, Dave. 1975. "Politics and Change in Dimam, Southwest Ethiopia". PhD Dissertation. The University of Kent.

Todd, D. 1977. "Caste in Africa" in *Africa* 47 (4):389-412.

Todd, D. "The origins of outcastes in Ethiopia: Reflections on an Evolutionary theory", in *Abbay* 9:145-158.

Todd, J. A. 1985. "Iron production by the Dimi of Ethiopia". Haaland, R. and Shinnie, P. (eds.) *African Iron Working: Ancient and Traditional.* Norwegian University Press, New York.

Todd, J.A. and Charles, J.A. 1978a. "Metallurgy as a contribution to archaeology in Ethiopia" in *Abbay* 9: 31-42.

Todd, J. and Charles, J. A. 1978b. "Ethiopian Bloomery Iron and the significance of inclusions Analysis in Iron Studies" in *Journal of the Historical Metallurgical Society* 12:63-87.

Tola, Genet. 1983. "From Ritual expert to petty chief: Qallu Abba Chafe of Aira" BA thesis. Addis Ababa University

Trigger, B. G. 2003. *Artifacts and Ideas: Essays in Archaeology.* Transaction Publishers. London.

Trigger, B.G. 1994. "Paradigms in Sudan Archaeology" In *International Journal of African Historical Studies*, 27 (2): 323-345.

Trigger, B. G.1969. "The Myth of Meroe and the African Iron Age" in *African Historical Studies,* II (I): 23-50.

Trimingham, J.S. 1952. *Islam in Ethiopia.*

Triulzi, Alessandro. 1996. "United and Divided: Borana and Gabaro among the Macha Oromo in Western Ethiopia" in P.T.W. Baxter, Jan Hultin and Alessandro Triulzi (Eds.) *Being and Becoming Oromo: Historical and Anthropological Enquiries.* Nordiska Afrikaninstitute, Uppsala.

Triulzi, Alessandro. 1986. "Nekemte and Addis Abeba: Dilemmas of provincialRule" in Donald Donham and Wendy James (eds.) *The Southern Marches of imperial Ethiopia: Essays in history and Social Anthropology.* Cambridge University Press.

Triulzi, Alessandro. 1981. *Salt, Gold and Legitimacy: Prelude to the history of a no-man's land Bela Shangul, Wallaga, Ethiopia (ca. 1800-1898).* Napoli, IUO.

Triulzi, Alessandro. 1975. "Trade, Islam, and the Mahdia in northwestern Wallaga" in *The Journal of African History*, Vol. 16, No.1.

Triulzi, Alessandro. 1973. "Some notes on the Galla countries south of the Abbay prior to the Battle of Embabo (1840s-1882)" Paper submitted to Historical Society of Ethiopia. A.A. (Memographed, Miscellaneous 1)

Turner, V. 1967. *The Forest of Symbols: Aspects of Ndembu Ritual.* Cornell University Press, ITHACA, New York.

Tylecote, R.F. 1980. "Furnaces, Crucibles, and Slags" in Theodore A. Wertime and James D. Muhly (eds.) *The Coming of the Age of Iron.* New Haven: Yale University Press.183- 228.

Tylecote, R.F. 1975. "The origin of iron smelting in Africa" in *West African Journal of Archaeology*, 5.

Ullendorff, E. 1960. *The Ethiopians: an introduction to country and people.* Lomdon, Oxford University Press.

Van Beek, Gus.W.1967. "Monuments of Axum in the light of South Arabian Archaeology" in *Journal of the American Oriental Society.* Vol.87, No.2.

Van der Merw, N.J. 1980. "The Advent of Iron in Africa" in Wertime and Muhly (eds.) *The Coming of the Age of Iron.* New Haven: Yale University Press. 463-506.

Van der Merwe, N.J. and Avery, D. 1987. "Science and Magic in African Technology: Traditional Iron Smelting in Malawi" in *Africa: Journal of the International African institute.* 57(2). Pp. 143-172.

Vansina, Jan. 2006. "Linguistic Evidence for the Introduction of ironworking into Bantu-speaking Africa" in *History in Africa*, Vol.33 (2006), pp. 321-361.

Vansina, Jan. 1995. "Historians, Are Archaeologists Your Siblings?" in *History in Africa*, Vol. 22(1995): 369-408. African Studies Association.

Wainwright, G.A.1942a. "Cosmas and the gold trade of Fazoqli" in *Man* 42:52-58. Wainwright, G.A. 1942b. "Early Records of iron in Abyssinia" in *Man.* 42:84-88. Wainwright, G.A. 1942C. "The coming of iron to some African peoples" in *Man.* 42:103-108. Waldbaum, J.C. 1980. "The First Archaeological appearance of iron and transition to the Iron Age" in Wertime and Muhly (eds.) *The coming of the Age of Iron.* New Haven: Yale University Press. 69-98.

Waqwaya, Chernet. 1988. "Land Tenure System and Self-settled Waloyes in Abe Dongoro Woreda : 1900-1974". BA Thesis. Addis Ababa University.

Wayessa, Bula S. 2010. 'Socialization, symbolism, and social structure: aspects of traditional pottery making among the Jimma Oromo, Western Oromia' in *Journal of Oromo Studies*, Vol. 17 (2) : 75-102.

Weedman, Kathryn. 2004 (Review of) *Ethnoarchaeology in Action* Nicholas David and Carol Kramer (2001) In *African Archaeological Review* Vol. 21 No.4, Dec. 2004.

Wertime, T.A. 1980. The Pyrotechnologic Background. In Wertime and Muhly (eds.) *The Coming of the Age of Iron.* New Haven: Yale University Press. 1-24.

Wilson, Lucy. 2007. "Terrain Difficulty as a factor in raw material procurement in the Middle Paleolithic of France" in *Journal of Field Archaeology*, Vol. 32 No.3, 2007.

Wolde-Mariam, Mesfin. 1972. *An Introductory Geography of Ethiopia.* Berhanena Selam H.S.I.

Printing Press, Addis Ababa.

Zewde, Bahru. 1987. "An Overview and Assessment of Gambella Trade (1904-1935)" in *The International Journal of African Studies*, Vol. 20, No.1.

Zewde, Bahru. 1991. *A History of Modern Ethiopia (18555-1974).* James Currey Ltd. London

9. LIST OF INFORMANTS

No.	Name of informants	Approximate age	Locality, zone or district	Period of interview	Remarks
1	Oliqaa Lamuu	c.65	Aira town	2007-2010	Son of renowned smelter from Laaloo lineage, vitally important information on locating former smelting sites in the present Aira town
2	Haayluu Amantee	c.85	Cudina Ayira	May 2009	Descendant of Wiinuu Qacoo, after whom the iron ore-mining site is called. Valuable informant on regional tradition.
3	Haayluu Gammadaa	c.80	Bonayya Bidiruu Aira	May 2009	Former smelter and resident close to ancestral smelting sites
4	Dojaa Horroo	c.100	Haroo Limmuu district, E. Wollega	July 2009	Belongs to former smelters/smiths. Despite his old age he never observed when smelting was active. Indicates that smelting has long abandoned.
5	Nagasuu Dayyaas	c.54	Haroo Limmuu	July 2009	He belongs to members of the 'Salgee', the nine councilors of Irreessa Qaallu site, participated in conflict resolution through metal tools
6	Waqrquu Galataa	c.89	>>	>>	>>
7	Yaadataa Soorii	c.70	>>	>>	>>
8	Tolasaa Garbaa	c.65	>>	>>	>>
9	Goobanaa Galataa	c.75	>>	>>	>>
10	Niguso Ciibsaa	c.76	>>	>>	Member of the ritual/ Qaalluu family of the Gibinaa lineage. Gibinaa ritual site is known as the place of 'unceasing fire'. This ritual site like the rest is associated with those who own the knowledge of metallurgy

11	Abdiisaa Nagarii	c.80	>>	>>	>>
12	Waaqtolee Dabalaa	65	>>	>>	Qaalluu head and smith. He presides over the Qaalluu site at Sagiroo Guddinaa. He and his councilors resolve conflicts in the medium of metal objects
13	Chernet Waqweyyaa	>50	Naqamte e	2005-2010 during fieldworks	Local historian, cultural expert, head of zonal culture office, native to Abee-Dongoroo area, first to refer to the presence of former smelters/ smiths in the region in his senior essay, collected oral tradition and was my vital field collaborator.
14	Moosisaa Gaaggaa	>90	Walagee, Abee-Dongoroo	June-November 2005	Malkee lineage-ore owners, miners, smelters. Extremely vital for the reenactment of smelting in 2005 but deceased in 2008.
15	Guddisaa Gaaggaa	c.90	>>	>>	Young brother of Moosisaa but with similar memory of the past activities around Faallee
16	Abdiisaa Miijanaa	c.82	>>	Sep. 2005	Malkee lineage and from senior line of those who controlled flow of resources, but were also smelters themselves. His mother had two smelting sites controlled by her dependant relatives.
17	Kabaa Miijanaa	c.66	>>	>>	Brother of Abdiisaa and provided information corroborating the formers' (nos. 14-16)
18	Nagawoo Qongaa	c.68	Sagiroo Gudina (Haroo Limmuu)	July 2009	Member of the'Salgee', the nine councilors of the Qaalluu site under Obbo Waaqtollee
19	Lammii Yaadataa	c.75	>>	>>	>>
20	Ejjetaa Nagarii	c.70	>>	>>	>>

21	Tafarraa Qana'aa	c.90	Aira/Jaar soo Abayyanii	Dec.2004, Sep-Nov, 2007, Dec.2010	Very important informant. Both smith and smelter. Cling to traditional belief of the Oromo. Master mind of the 2007 smelting reconstruction
22	Shalbee Guutamaa	c.90	>>	Decemb. 2004	A close friend of Obbo Tafarraa. Similar experience with him. But passed away in Sep.2007
23	Waratii Fayyisaa	c.87	Yubdo district	November 2007	Furnace builder and tuères maker. He did this for one of the two sites in 2007 and demonstrated for the young. Very important on skills and less on story.
24	Ofga'aa Tuuchoo	>90	Aira town	May 2009	He departed from other informants on the way furnace used to be built. He remembers furnace without wooden wall. A very good insight on variations of local technology.
25	Kumarraa Sanbatoo	c.65	Abayyan ii	Dec. 2004, 2007, 2009, 2010	He belongs to the lineage of Tafarraa Qana'aa. Obbo Tafarra entrusted the organizing role for the 2007 smelting on him. He is a well known smith and well off.
26	Yaadataa Guutaa	>95	Abayyan ii	Dec.2010	A respected oral informant. But due to age and afraid to discuss in front of artisans, he could not provide much valuable information.

27	Ejjetaa Joobir	c.75	Aira	Octo. 2007	Observed former smelting and was enlightened protestant Christian. He was good at genealogies.
28	Nagash Fayyisaa	c.65	Walagee	June-October 2005	Son of an ex-slave . A living testimony that the smiths were able to have slaves in the past. He guided the young re-enactor in ore processing and furnace construction.
29	Darbuush Kumsaa	c.48	Walagee	>>	Son of a known smith and the family of chief of smiths. He volunteered to demonstrate smelting in 2005 with the help of Nagash and guidance from former smelters such as Moosisaa and Guddisaa.